Molecular Biology
Biochemistry and Biophysics
30

David Vázquez

Inhibitors of Protein Biosynthesis

With 61 Figures

Springer-Verlag Berlin Heidelberg New York 1979

Dr. David Vázquez
Instituto de Bioquimica de Macromoléculas
Centro de Biologia Molecular
C.S.I.C. and U.A.M. Canto Blanco
Madrid 34/Spain

ISBN-13:978-3-642-81311-5 e-ISBN-13:978-3-642-81309-2
DOI: 10.1007/978-3-642-81309-2

Library of Congress Cataloging in Publication Data. Vázquez, David. Inhibitors of protein biosynthesis. (Molecular biology, biochemistry, and biophysics; 30) Bibliography: p. Includes index. 1. Protein biosynthesis. 2. Genetic translation. I. Title. II. Series. QP551.V42 574.1'9296 78-26030.

Softcover reprint of the hardcover 1st edition 1979

2131/3130-543210

Contents

List of Abbreviations

cAMP	Cyclic adenosin-monophosphate
RNA	Ribonucleic acid
mRNA	Messenger ribonucleic acid
MDMP	2-(4-methyl-2,6-dinitroanilino)-N-methyl-propionamide
-2'(3'),5'-ADP	2'(3'),5'-Adenosintriphosphate
PAP	Phytolacca americana protein
TPCK	1-Chloro-4-phenyl-3-tosylamido-2-butanone
PA toxin	Pseudomonas aeruginosa toxin
f-Met-$tRNA_F$ or fMet-$tRNA_F$	Initiator formyl-methionyl-tRNA of prokaryotes
Met-$tRNA_F$	Initiator methionyl-tRNA of eukaryotes
GTP	Guanosintriphosphate
IF-1	Bacterial initiation factor 1
IF-2	Bacterial initiation factor 2
IF-3	Bacterial initiation factor 3
eIF-1	Eukaryotic initiation factor 1
eIF-2	Eukaryotic initiation factor 2
eIF-3	Eukaryotic initiation factor 3
eIF-4A	Eukaryotic initiation factor 4A
eIF-4B	Eukaryotic initiation factor 4B
eIF-4C	Eukaroytic initiation factor 4C
eIF-5	Eukaroytic initiation factor 5
$m^7G(5')ppp$ or $m^7G^{5'}ppp$	"cap" or 7-methylguanosine-5'-triphosphate
SV40	Simian virus 40
$m^7G^{5'}p$	7-Methylguanosine-5'-monophosphate
tRNA	Transfer ribonucleic acid
rRNA	Ribosomal ribonucleic acid
EF-2	Elongation factor 2

ATP	Adenosin-triphosphate
Ac-Phe	Acetyl-phenylalanine
EF-G	Elongation factor G
GTPase	GTPasa
EF-Ts	Elongation factor Ts
Poly(U)	Polyuridilyc acid
DNA	Deoxyribonucleic acid
$m^7G^{5'}pp$	7-Methylguanosine-5'-diphosphate
AMP	Adenosin-monophosphate
GMPPCP or GDPCP	Guanylyl methylenediphosphonate
GDPNP	Guanylyl imidodiphosphate
EF-T	Elongation factor T
RF-1	Release factor 1
RF-2	Release factor 2
RF-3	Release factor 3
m^7G	7-Methylguanosine
$5'pm^7G$	5'-Phosphate-7-methylguanosine
Sm^D	Streptomycin dependent
EF-Tu	Elongation factor Tu
EF-1	Elongation factor 1
C. diphtheriae	*Corynebacterium diphtheriae*
NAD	Nicotinamide-adenine dinucleotide
PRT	*Penicillium roqueforti* toxin

Chapter 1
Protein Synthesis and Translation Inhibitors

1. Introduction

Studies concerning selectivity, site,and mode of action of translation inhibitors have been widely developed and the subject has been reviewed repeatedly in the last fifteen years (Gale, 1963; Newton, 1965; Newton and Reynolds, 1966; Gottlieb and Shaw, 1967a; Vãzquez and Monro, 1967; Weisblum and Davies, 1968; Pestka, 1971; Muñoz, García-Ferrandiz and Vãzquez, 1972; Gale et al., 1972; Kaji, 1973; Vãzquez, 1974; Corcoran and Hahn, 1975; Pestka, 1977; Vãzquez, 1978a). This study will be concerned mainly with the present state of the problem. A complete survey of the literature will not be possible in this contribution and the above reviews should be very useful to readers interested in different aspects of the problem. Furthermore, there are a number of reviews concerning the chemistry, biosynthesis, toxicology, inhibitory spectra of the different inhibitors (Korzybski, Kowszyk-Gindifer and Kurylowicz, 1967a, 1967b; Gottlieb and Shaw, 1967a, 1967b; Glasby, 1976) and mechanism of drug resistance (Benveniste and Davies, 1973a; Mitsuhashi, Rosival and Krčméry, 1975; Mitsuhashi, 1977; Mitsuhashi and Hashimoto, 1977) which should be very useful to readers who are concerned with these topics.

Specific inhibitory effects on bacterial protein synthesis by chloramphenicol and chlortetracycline, at their minimal growth inhibitory concentrations, were first described by Gale and Paine (1950a, 1950b). At those concentrations the antibiotics did not affect respiration, fermentation,or amino acid accumulation (Gale and Paine, 1950b), but caused an inmediate cessation of protein synthesis and an increase in the rate of nucleic acid accumulation in bacteria (Gale and Folkes, 1953). Similar effects were later observed in bacteria treated with a number of translation inhibitors (Gale, 1963; Newton, 1965; Newton and Reynolds, 1966; Gottlieb and Shaw, 1967a; reviews). The mechanism of protein synthesis remained obscure during the period 1950-1960, so that it was not possible to establish the site of action of the known translation inhibitors. A resolved cell-free system to study ribosomal amino acid incorporation directed by a synthetic polynucleotide such as mRNA was first described in 1961,and the specific inhibitory effect of chloramphenicol was confirmed in this system (Nirenberg and Matthaei, 1961). Model systems to study the individual reactions in protein synthesis were further developed in the following years and thus the specific steps blocked by different inhibitors of translation were elucidated.

Reports concerning inhibition of translation in higher cells did not appear until 1958,when the inhibitory effect of cycloheximide on protein synthesis in *Saccharomyces carlsbergensis* was described. Contrary to what was found in chloramphenicol-treated bacteria, the stringent control of nucleic acid synthesis in yeast was not abolished by cycloheximide and an inhibition of nucleic acid synthesis was also observed in the presence of the antibiotic (Kerridge, 1958). Similar results were later observed in higher cells treated with other translation inhibitors (Newton and Reynolds, 1966; Gottlieb and Shaw, 1967a; reviews). Cell-free systems and model reactions to study protein synthesis have been developed in the last twenty years and the reactions blocked by the different inhibitors were elucidated.

For the sake of clarity we have adopted for the protein factors involved in translation the nomenclature developed at the International Symposia on Protein Synthesis held at the National Institutes of Health (Bethesda, USA) in 1971 and 1976 (Anderson et al., 1977). The nomenclature for the proteins of the bacterial and mammalian ribosomes is adopted from early studies on the subject (Kaltschmidt and Wittmann, 1970; Sherton and Wool, 1972).

2. Site of Action of Protein Synthesis Inhibitors

The process of protein synthesis can be arbitrarily divided into (a) steps taking place in early reactions in protein synthesis and (b) steps in the translation mechanisms taking place at the ribosome level. Following these criteria protein synthesis inhibitors of group (a) can be classified as indicated in Table 1. However, considering the specificity, selectivity, and permeability of the protein synthesis inhibitors, the most important compounds are undoubtedly those included within the ample group of translation inhibitors (see Tables 2, 3, and 4). Therefore we will now refer specifically to this group. The overall reactions inhibited by these compounds are indicated in Figures 1 and 2. However a number of inhibitors have pleiotropic effects on the ribosome and inhibit more than one reaction; in these cases we usually indicate the principal step(s) blocked by the inhibitors. A number of inhibitors are not presented in

Table 1. Inhibitors of protein synthesis acting on steps taking place prior to translation

Inhibitors of aminoacyl-tRNA formation		
Inhibitors of amino acid activation[a]	Inhibitors of amino acid transfer to RNA[a]	Inhibitors which are transferred to tRNA leading to synthesis of abnormal proteins[a]
7-Azatryptophan	Aminoalkyl-adenylates	Ethionine
Tryptazan	Borrelidin[b]	Norleucine
6-Fluorotryptophan	Furanomycin[b]	Alloisoleucine
5-Fluorotryptophan	Minosine[b]	Azetidine-carboxylic acid
Norvaline	4-Oxalysine	Canavanine
α-Amino-β-clorobutyrate	2,6-Diamino-4-hexynoic acid	N-Ethylglycine
α-Aminobutyrate	Trans-4-dehydrolysine	O-Methyl threonine
Selenomethionine		2-Fluoro-L-histidine[c]
Ethionine		
Norleucine		
Methyl-ester of serine		
Ethyl-ester of serine		
Aminoalkyl-adenylates		
Tiramine		
L-Tyrosinol		
L-Tyrosine amide		
L-Tyrosine methyl ester		

[a]Amino acid analogs specifically replace or compete with their corresponding amino acids.

[b]Borrelidin, furanomycin,and minosine specifically inhibit threonyl-, isoleucyl-, and phenylalanyl-tRNA synthesis respectively.

[c]Klein et al. (1977). Other data are taken from Vâzquez (1974; review) and references therein.

Table 1 (continued)

Inhibitors of f-Met-tRNA$_F$ formation		
Inhibitors of N^{10}-formyl-H_4 folate synthesis	Inhibitors depleting the pool of N^{10}-formyl-H_4 folate	Analogs of N^{10}-formyl-H_4 folate
Aminopterin Amethopterin (synonym methotrexate) Pteroylaspartic acid Trimethoprim 6-Chloro-8-aza-9-cyclopentylpurine	Hydroxylamine	Pyrimidine analogs Pteridine analogs: tetrahydropteroate N^5-formyl-H_4 folate N^5-methyl-H_4 folate tetrahydrohomofolate tetrahydrohomopteroate

Table 2. Inhibitors of translation acting on prokaryotic systems

- Althiomycin
- Avilamycin
- Berninamycin
- Bottromycin A_2
- Chloramphenicol group:
  - Chloramphenicol
  - D-AMP-3
  - D-Thiomycetin
  - D-Win-5094
- Cloacin DF13
- Colicin E3
- Griseoviridin
- Kasugamycin
- Lincomycin group:
  - Celesticetin
  - Clindamycin
  - Lincomycin
- Macrolide antibiotics:
  - Carbomycin group:
    - Carbomycins
    - Josamycin
    - Leucomycins
    - Niddamycins
  - Erythromycin group:
    - Erythromycins
    - Neospiramycins
    - Oleandomycin
  - Lancamycin group:
    - Chalcomycin
    - Kujimycin A
    - Lancamycin
  - Methymycin group:
    - Forocidins
    - Methymycin
    - Narbomycin
    - Neomethymycin
    - Picromycin
- Micrococcin
- Negamycin
- Rubradirin
- Spectinomycin
- Streptogramin A group:
  - Ostreogrycin G
  - Streptogramin A
- Streptogramin B group:
  - Staphylomycin S
  - Streptogramin B
  - Viridogrisein
- Streptomycin group:
  - Amikamycin
  - Gentamicin
  - Kanamycin
  - Neomycin
  - Paromomycin
  - Sisomicin
  - Streptomycin
  - Tobramycin
- Streptothricins
- Thermorubin
- Thiostrepton group:
  - Siomycin
  - Sporangiomycin
  - Thiopeptin
  - Thiostrepton
- Viomycin group:
  - Capreomycin
  - Viomycin

Table 2 (continued)

- Spiramycin group:
  - Angolamycin
  - Relomycin
  - Spiramycins
  - Tylosin

Table 3. Inhibitors of translation acting on eukaryotic systems

- Abrin
- Alpha sarcin
- Anisomycin
- 5-Azacytidine
- Bruceantin
- Crotins
- Curcins
- Diphtheria toxin
- Emetine group:
  - Emetine
  - Tubulosine
- Enomycin
- Glutarimide group:
  - Actiphenol
  - Cycloheximide
  - Streptimidone
  - Streptovitacin A
- Harringtonine group:
  - Harringtonine
  - Homoharringtonine
  - Isoharringtonine
- Lycorine group:
  - Lycorine
  - Pseudolycorine
- MDMP
- Narciclasine group:
  - Haemanthamine
  - Narciclasine
  - Pretazettine
- PAP
- Pederine
- Phenomycin
- Ricin
- Sodium fluoride
- Tenuazonic acid
- Trichotecene antibiotics:
  - Trichodermin group:
    - Fusarenon-X
    - Trichodermin
    - Trichodermol
    - Trichothecin
  - Verrucarin A group:
    - Deacetoxyscirpenol
    - Nivalenol
    - Toxin T-2
    - Verrucarin A
- Tylophora alkaloids:
  - Cryptopleurine
  - Tylocrebrine
  - Tylophorine

Table 4. Inhibitors of translation acting on prokaryotic and eukaryotic systems

Actinobolin	Hygromycin B
Adrenochrome	Nucleocidin
AHR-1911	Pactamycin
Amicetin group:	Polydextran sulphate
Amicetin	Polyvinyl sulphate
Bamicetin	Puromycin
Plicamicetin	Pyrochatechol violet
Anthelmycin	Showdomycin
Aurintricarboxylic acid	Sparsomycin
Blasticidin S	Tetracycline group:
Chartreusin	Chlortetracycline
Edeine A_1	Doxycycline
Fusidic acid	Oxytetracycline
Gougerotin	Tetracycline
Guanylyl-methylene-diphosphate	Tosylphenylalanylchloromethane
Guanylyl-imido-diphosphate	

Fig. 1. Translation process in bacteria. Site of action of translation inhibitors. ▶
* Do not interact with polysomes. Therefore bind to free ribosome subunits and prevent only the first few rounds of peptide bond formation. ** Is an inhibitor of aminoacyl-tRNA binding in intact cells or in integrated systems in which elongation is proceeding in the presence of EF-Tu and EF-G. Does not inhibit aminoacyl-tRNA binding in resolved systems in the absence of EF-G. Can inhibit translocation in cell-free systems. *** Can also inhibit translocation in cell-free systems. **** Does not inhibit peptide bond formation in resolved assays. However, it blocks this step in intact cells and integrated systems by preventing the release of EF-Tu·GDP bound to the ribosome. ***** Do not interact with polysomes. Do not inhibit translocation in many model systems. Bind to free ribosome subunits and prevent elongation of the nascent polypeptide chain when it reaches a certain size, before polysome formation

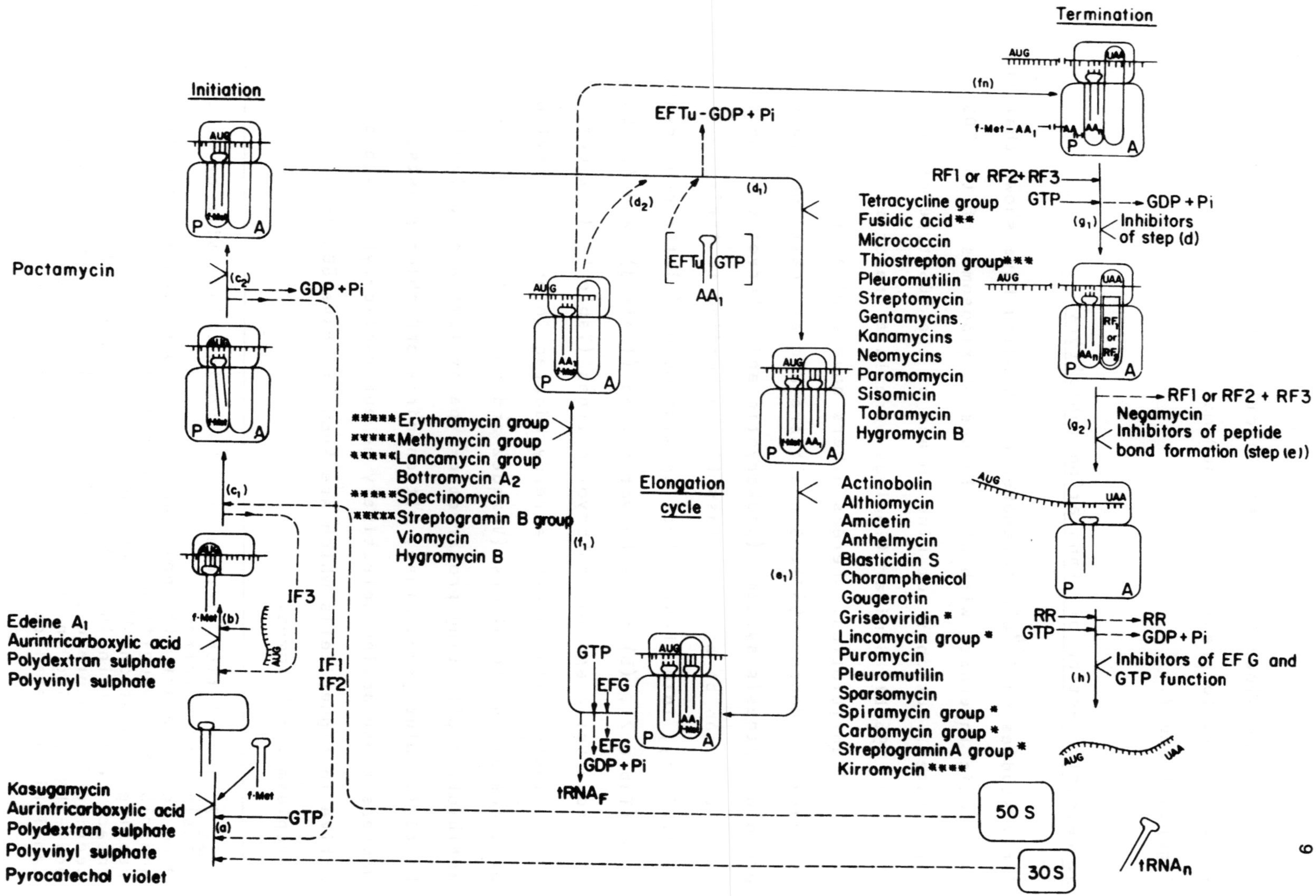
Initiation
Elongation cycle
Termination
Pactamycin
Edeine A1
Aurintricarboxylic acid
Polydextran sulphate
Polyvinyl sulphate
Kasugamycin
Aurintricarboxylic acid
Polydextran sulphate
Polyvinyl sulphate
Pyrocatechol violet
(a)
(b)
(c1)
(c2)
GDP + Pi
GTP
IF3
IF1
IF2
f-Met
AUG
P
A
EFTu - GDP + Pi
EFTu GTP
AA1
(d1)
(d2)
(e1)
(f1)
(fn)
*****Erythromycin group
*****Methymycin group
*****Lancamycin group
Bottromycin A2
*****Spectinomycin
*****Streptogramin B group
Viomycin
Hygromycin B
GTP
EFG
EFG
GDP + Pi
tRNAF
Tetracycline group
Fusidic acid**
Micrococcin
Thiostrepton group***
Pleuromutilin
Streptomycin
Gentamycins
Kanamycins
Neomycins
Paromomycin
Sisomicin
Tobramycin
Hygromycin B
Actinobolin
Althiomycin
Amicetin
Anthelmycin
Blasticidin S
Choramphenicol
Gougerotin
Griseoviridin *
Lincomycin group *
Puromycin
Pleuromutilin
Sparsomycin
Spiramycin group *
Carbomycin group *
Streptogramin A group *
Kirromycin ****
f-Met - AA1
RF1 or RF2+RF3
GTP
GDP + Pi
(g1)
Inhibitors of step (d)
UAA
RF1 or RF2 + RF3
(g2)
Negamycin
Inhibitors of peptide bond formation (step (e))
RR
GTP
RR
GDP + Pi
(h)
Inhibitors of EF G and GTP function
50 S
30 S
tRNAn

these figures since their sites of action are not well known or are difficult to present in the schematic representation of Figures 1 and 2.

3. Selective Action of Translation Inhibitors

Early studies on (^{14}C)chloramphenicol binding have shown that the antibiotic interacts with all classes of ribosomes of the 70S type tested, but does not bind to any of the 80S type ribosomes (Vázquez, 1964b). Similar findings were later reported for a number of antibiotics, whereas others have a wider spectrum (Vázquez and Monro, 1967). Since there are at least two types of protein synthesis system (prokaryotic and eukaryotic), their inhibitors can be broadly classified according to their specificity, into those affecting systems of (a) the prokaryotic type (Table 2), (b) the eukaryotic type (Table 3), and (c) both the prokaryotic and the eukaryotic types (Table 4) (Vázquez, 1974). These criteria were already adopted in an early review of translation inhibitors (Vázquez and Monro, 1967). In general, antibiotics affecting prokaryotic-type systems are active in bacteria, blue-green algae, mitochondria and chloroplasts, whereas those acting selectively on eukaryotic-type systems are active in higher cells which are known to have 80S type cytoplasmic ribosomes.

Fig. 2. Translation process in eukaryotic cells. Site of action of translation inhibitors. ▶
* Do not interact with polysomes. Therefore bind to free ribosomes and ribosome subunits and prevent only the first few rounds of peptide elongation. ** Is an inhibitor of aminoacyl-tRNA binding in intact cells or in integrated systems in which elongation is proceeding in the presence of EF-1 and EF-2. Does not inhibit aminoacyl-tRNA binding in resolved systems in the absence of EF-2. Can inhibit translocation in cell-free systems

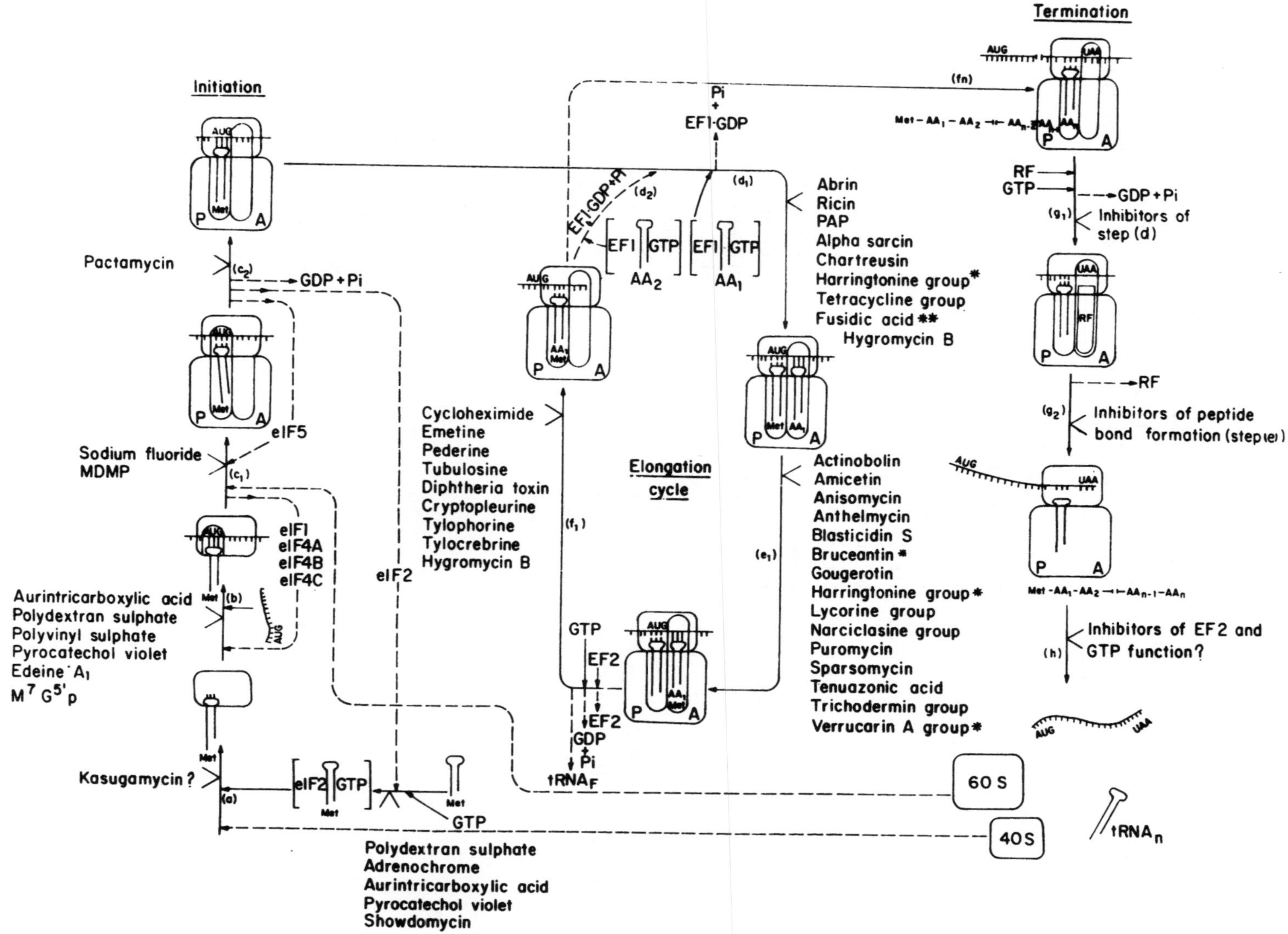

Initiation
Elongation cycle
Termination
Pactamycin
GDP + Pi
eIF5
Sodium fluoride
MDMP
eIF1
eIF4A
eIF4B
eIF4C
Aurintricarboxylic acid
Polydextran sulphate
Polyvinyl sulphate
Pyrocatechol violet
Edeine A1
M7 G5' p
Kasugamycin ?
eIF2
GTP
Met
Polydextran sulphate
Adrenochrome
Aurintricarboxylic acid
Pyrocatechol violet
Showdomycin
Cycloheximide
Emetine
Pederine
Tubulosine
Diphtheria toxin
Cryptopleurine
Tylophorine
Tylocrebrine
Hygromycin B
EF1·GDP+Pi
Pi + EF1·GDP
EF1 GTP AA2
EF1 GTP AA1
EF2
GDP + Pi
tRNAF
Abrin
Ricin
PAP
Alpha sarcin
Chartreusin
Harringtonine group*
Tetracycline group
Fusidic acid **
Hygromycin B
Actinobolin
Amicetin
Anisomycin
Anthelmycin
Blasticidin S
Bruceantin *
Gougerotin
Harringtonine group *
Lycorine group
Narciclasine group
Puromycin
Sparsomycin
Tenuazonic acid
Trichodermin group
Verrucarin A group *
Met – AA1 – AA2 – AAn
RF
GTP
Inhibitors of step (d)
Inhibitors of peptide bond formation (step (e))
Inhibitors of EF2 and GTP function?
60 S
40 S
tRNAn

Table 5. Site of action of translation inhibitors

Other cell fractions (site of action or component with which it interferes)	Ribosome	
	Small ribosomal subunit (30S* or 40S** or both***)	Larger ribosomal subunit (50S* or 60** or both***)
Fusidic acid (EF-G and EF-2)	Adrenochrome**	Abrin**
Kirromycin (EF-Tu)	Aminoglycoside antibiotics*	Actinobolin***
Diphtheria toxin (EF-2)	Aurintricarboxylic acid***	Alpha sarcin**
PA toxin (EF-2)	Chartreusin***	Althiomycin*
Guanylyl-methylene-diphosphate	Cloacin DF13*	Amicetin group***
(GTP)	Colicin E3*	Anisomycin**
Guanylyl-imido-diphosphate	Edeine A_1***	Anthelmycin***
(GTP)	Kasugamycin***	Blasticidin S***
Hemin-controlled repressor	MDMP**	Bottromycin A_2*
(eIF-2)	Pactamycin***	Bruceantin**
cAMP (cAMP-dependent protein	Polydextran sulphate***	Chloramphenicol group*
kinase)	Polyvinyl sulphate***	Crotins**
Double-stranded RNA	Sodium fluoride**	Curcins**
Showdomycin (eIF-2 and EF-2)	Spectinomycin group*	Emetine group**
Interferon	Streptomycin*	Enomycin**
	Tylophora alkaloids**	Glutarimide group**
		Gougerotin***
		Griseoviridin*
		Harringtonine group**
		Lincomycin group*
		Lycorine group**
		Macrolide antibiotics*
		Micrococcin*
		Multhiomycin*
		Narciclasine group**
		Negamycin*
		Nucleocidin***
		PAP**
		Phenomycin**
		Puromycin***
		Ricin**

Table 5 (continued)

Other cell fractions (site of action or component with which it interferes)	Ribosome	
	Small ribosomal subunit (30S* or 40S** or both***)	Larger ribosomal subunit (50S* or 60** or both***)
		Sparsomycin***
		Streptogramin A group*
		Streptogramin B group*
		Thiostrepton group*

In most cases it is already established that the specific inhibitory effect of a translation inhibitor is a consequence of a specific stoichiometric interaction with or enzymic action of the inhibitor on a particular structure or component involved in some step of translation. According to these criteria we can classify inhibitors of translation as indicated in Table 5, considering in all cases the specificity shown above (Tables 2-4).

4. Translation of mRNA

The overall reactions taking place in the process of translation by prokaryotic and eukaryotic ribosomes can be represented schematically according to the translocation model with one or two entry sites on the small subunit. Following this model, the process of translation can be arbitrarily divided into three phases: initiation, elongation, and termination (Figs. 1 and 2). Our knowledge of the translational steps and the specific effects of the various inhibitors has progressed in parallel over the last fifteen years. The most recent developments concerning the mechanism of translation and the specific requirements for the different steps have been extensively reviewed (Cold Spring Harbor Symp.Quant.Biol., 1969; Lucas-Lenard and Lipmann, 1971; Bosch, 1972; Haselkorn and Rothman-Denes, 1973; Jackson, 1975; Tate and Caskey, 1974; Modolell and Vázquez, 1975; Chapeville and Grunberg-Manago, 1975; Weissbach and Pestka, 1977). Consequently we will refer only briefly to these topics, describing more extensively the site of action of the inhibitors as shown schematically in Figures 1 and 2.

Chapter 2

Initiation

1. Introduction

Both ribosomal subunits, the initiator substrate (f-Met-$tRNA_F$ in bacteria and Met-$tRNA_F$ in eukaryotes), GTP, mRNA and a number of protein factors (IF-1, IF-2 and IF-3 in bacteria; eIF-1, eIF-2, eIF-3, eIF-4A, eIF-4B, eIF-4C and eIF-5 in eukaryotes) are required in the steps of initiation of translation. There is only one initiation phase per protein chain synthesized in which we have to distinguish at least three steps: (a) recognition of initiation factors and binding of the initiator substrate (f-Met-$tRNA_F$ in prokaryotes and Met-$tRNA_F$ in eukaryotes), (b) recognition of the mRNA initiation triplet and (c) the final steps of initiation involved in the joining of the complex to the larger ribosome subunit (Figs. 1 and 2). Step (c) involves the reactions (c_1), in which the larger subunit is attached to the initiation complex, and (c_2) in which the 3' end of the initiator substrate binds to the donor site of the peptidyl-transferase center. According to the model for one single-site interaction on the small subunit the anticodon interaction of the initiator with the small subunit,must be interrupted in step (c_2), leaving the site open for the entry of aminoacyl-tRNA to start the elongation phase (Figs. 1 and 2). The interaction is however continued, according to the two entry sites on the small ribosomal subunit.

It is widely, but not wholly accepted that step (a) precedes step (b) in eukaryotic systems, and it is still a matter of controversy whether this is so in bacterial systems. However for comparative purposes and to favor conciseness we will consider here that the sequence of the reactions in initiation follows the steps (a)→(b)→(c) in both prokaryotic (Fig. 1) and eukaryotic (Fig. 2) systems.

An important finding has recently been described in the nucleotide moiety at the 5' end of some mRNA species. The methylated terminal $m^7G(5')ppp$.... ("cap") has been observed at the 5' end of mRNA isolated from mouse, monkey, and human cells as well as in reovirus, vesicular stomatitis virus, vaccinia virus, and SV40 (Both et al., 1975). In these mRNA the $m^7G^{5'}p$ at the 5' end appears to be essential for translation and is specifically required for step (b) (Both et al., 1975) in a reaction where eIF-4B appears to be involved in the recognition of this 5' end moiety and eIF-4C recognizes other part(s) of the mRNA. By using vesicular stomatitis virus mRNAs with these modified 5' termini, translation was studied in wheat germ and reticulocyte cell-free protein synthesis systems; the 5'-terminal $m^7G^{5'}p$ was shown to be quantitatively more important in the heterologous plant system than in the reticulocyte lysate (Lodish and Rose, 1977). The presence of this $m^7G^{5'}p$ has not been shown in mRNA from bacteria, satellite tobacco necrosis virus, or encephalomyocarditis virus and does not appear to be required for translation in these cases.

Translation inhibitors classified within the group specifically blocking initiation can be characterized in cell-free systems for translation by the following criteria altogether: (a) They preferentially inhibit natural mRNA-directed synthesis of polypeptides and hardly affect synthetic polynucleotide-directed amino acid incorporation without physiological initiation. (b) When added to a system actively synthesizing polypeptides,they allow polypeptide synthesis of chains which have already initiated and therefore, in this system, there is a time lag before inhibition of synthesis is detected,i.e.,until new initiation takes place and is inhibited. (c) They cause polysome breakdown in intact cells or cell-free systems which have already initiated,since they allow polysome run-off and prevent polysome formation. However polysome breakdown in itself is not sufficient to class a drug as an inhibitor of initiation, as has been stated in some cases; indeed there are a number of elongation inhibitors that also cause polysome breakdown since they do not interact with polysomes and only block the first few rounds of elongation prior to polysome formation. This is the case with lincomycin, erythromycin, and antibiotics of the streptogramin A and streptogramin B groups in bacteria; similarly, verrucarin A,harringtonine, and bruceantin are elongation inhibitors which cause polysome run-off in eukaryotes since they only block the first few rounds of elongation and are not active on polysomes.

The individual steps taking place in the process of initiation are usually studied in resolved assays using the purified initiation factors and washed ribosomal subunits derived from the total cellular ribosomes and polysomes. Most of the present conclusions concerning the specific effects of initiation inhibitors are based on this type of assay, and therefore come in for some criticism. Thus the inhibitory effects of some drugs are not shown since the reactions in these resolved systems are due to stoichiometric interactions (initiation factor and GTP with the ribosomal subunits). Some drugs, such as pactamycin, on the other hand, appear to interfere with the catalytic reactions and recycling of some components which take place only in the normal process of translation in intact cells or cell-free systems. Furthermore, in the case of eukaryotes, the results obtained from testing numerous initiation inhibitors are different when either derived 40S subunits (prepared from ribosomes and polysomes) or native 40S subunits (complexes present in cells containing 40S ribosomes with different initiation factors bound)are used . These native 40S subunits certainly appear to be intermediate complexes which have already initiated the process of translation and therefore the effects of the inhibitors of steps (a) and/or (b) are very small or nil when native 40S subunits are used in resolved systems to test the inhibitors (i.e., edeine A_1, adrenochrome, and aurintricarboxylic acid) (Fresno and Vázquez, 1978).

2. Inhibitors of Recognition of the Initiator Substrate [Step (a)]

This step is not well resolved in bacterial systems. In eukaryotes it involves at least two reactions: in the first one formation of the eIF-2-Met-$tRNA_F$-GTP complex takes place, and in the second one this complex is attached to the ribosome. Therefore, we have some inhibitors that prevent formation of the eIF-2-Met-$tRNA_F$-GTP complex in eukaryotes (showdomycin) whereas some others interact with the ribosome preventing the attachment of the initiator substrate fMet-$tRNA_F$ in bacteria or the eIF-2-Met-$tRNA_F$-GTP complex (kasugamycin) in eukaryotes.

2.1 Kasugamycin

Kasugamycin hydrochloride has been crystallized from culture filtrates of a strain of Streptomyces kasugaensis, which were very active in preventing rice blast by the fungus Piricularia oryzae (Umezawa et al., 1965). Indeed kasugamycin has a strong antifungal effect at rather low pH (4.0 to 5.0) and is also active on gram-positive and gram-negative bacteria (Hamada et al., 1965). The structure of the aminoglycoside antibiotic kasugamycin was elucidated early on (Fig. 3) (Ikekawa, Umezawa and Sitaka, 1966).

Kasugamycin stoichiometrically interacts with the 30S ribosomal subunit (Okuyama and Tanaka, 1975) and blocks translation in prokaryotic (Tanaka, Yamaguchi and Umezawa, 1966) and fungal systems (Masukawa, Tanaka and Umezawa, 1968). However kasugamycin does not cause misreading (Tanaka, Yamaguchi and Umezawa, 1966; Masukawa, Tanaka and Umezawa, 1968) since it does not posses the characteristic deoxystreptamine or streptamine moiety common to all aminoglycoside antibiotics which do cause misreading. Early reports have described a certain inhibitory effect of kasugamycin on polynucleotide-directed aminoacyl-tRNA binding to bacterial (Tanaka, Yamaguchi and Umezawa, 1966) and

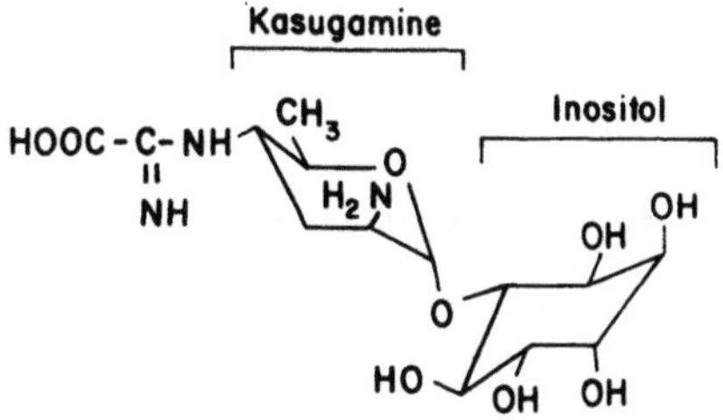

$C_{14}H_{25}O_9N_3$; MW 379.36

Fig. 3. Chemical structure of kasugamycin

fungal (Masukawa, Tanaka and Umezawa, 1968) ribosomes. However, further work has shown that the antibacterial effect of the antibiotic is due to its inhibitory effect on initiation. Indeed kasugamycin inhibits the binding of f-Met-$tRNA_F$ to the 30S ribosomal subunit (Okuyama et al., 1971; Okuyama, Watanabe and Tanaka, 1972) and thus blocks the translation of f2 phage RNA at the level of the initiation phase (Okuyama and Tanaka, 1972; Kozak and Nathans, 1972; Okuyama and Tanaka, 1973). A mutation altering the 30S ribosomal subunit and giving rise to resistance to kasugamycin was early observed in E. coli and mapped near the leucine region (therefore far from the streptomycin region) (Sparling, 1970). Resistance to kasugamycin was reported to be due to a lack of methylation of two adjacent adenine residues in the 16S rRNA of the 30S subunit (Helser, Davies and Dahlberg, 1971). It was further observed that sensitive bacteria contain an RNA methylase activity that dimethylates in vitro the relevant adenine residues of 16S rRNA of the resistant mutant, and that the 30S subunits reconstituted with this 16S rRNA methylated in vitro are sensitive to the antibiotic (Helser, Davies and Dahlberg, 1972). Although an alteration of protein S4 was observed in some kasugamycin-resistant mutants it was due to cotransduction of both loci, whereas ribosomal resistance to the antibiotic was clearly due to 16S rRNA alteration and not to any change in protein S4 (Zimmermann, Ikeya and Sparling, 1973). However, another type of E. coli mutant has been isolated in which the resistance to kasugamycin is attributed to an alteration of ribosomal protein S2 (Okuyama, Yoshikawa and Tanaka, 1974). Although the effects of kasugamycin on initiation in bacterial systems are clearly documented, similar studies in fungi systems have not been performed, although the antibiotic might be expected to act similarly in bacteria and fungi.

2.2 Showdomycin

The antibiotic showdomycin was crystallized from culture filtrates of Streptomyces showdoensis (Nishimura et al., 1964). It is a very wide spectrum antibiotic since it is not only active in gram-positive and gram-negative bacteria but also in eukaryotic and tumoral cells (Nishimura et al., 1964). It is an analog of uridine and of maleimide (Fig. 4) (Suhadolnik, 1970) and therefore has a very complex action in bacteria and eukaryotic cells. Showdomycin, owing to its analogy with uridine, inhibits certain enzymes involved in the metabolism of pyrimidines (Suhadolnik, 1970) and owing to its analogy with maleimide blocks reactions in which -SH groups are involved. Thus in the process of translation showdomycin inhibits initiation by preventing formation of the eIF-2-Met-$tRNA_F$-GTP complex (Fresno, Carrasco and Vázquez, 1976) and elongation by interfering with the function

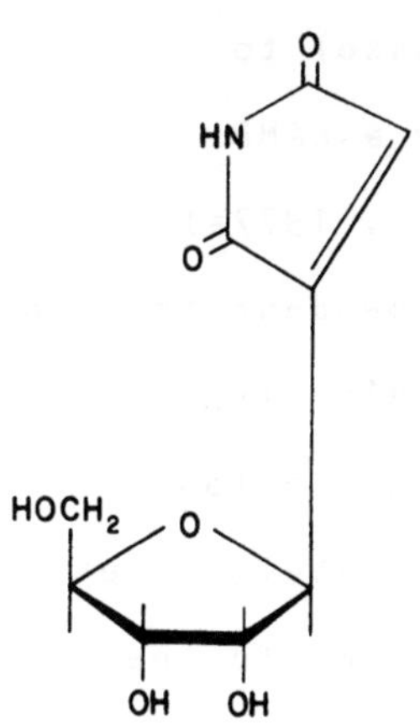

$C_9H_{11}O_6N$; MW 229.19

Fig. 4. Chemical structure of showdomycin

of EF-2 on the 60S ribosomal subunit (Bermek et al., 1970; Bermek, Monkemeyer and Berg, 1971). Furthermore showdomycin inhibits the DNA polymerase II (Sung, 1972) and the RNA polymerase (Maryanka and Johnston, 1970) in E. coli.

2.3 Hemin-Controlled Repressor (Translational Inhibitor) and cAMP

In reticulocyte lysates hemin prevents the formation of an inhibitor (MW 300,000) of chain initiation from a proinhibitor of similar molecular weight (Gross and Rabinovitz, 1972). This translational inhibitor (hemin-controlled repressor) has no effect on formation of the eIF-2-Met-$tRNA_F$-GTP complex but prevents Met-$tRNA_F$-GTP complex binding to the 40S ribosomal subunit; this inhibitory effect can be overcome in reticulocyte lysates by addition of either hemin or eIF-2 (Clemens et al., 1974; Clemens, 1976). The inhibitor is a cAMP-independent protein kinase that catalyzes the phosphorylation of the small subunit (MW 38,000) of eIF-2 (Kramer, Cimadevilla and Hardesty, 1976; Pinphanichakarn, Kramer and Hardesty, 1976; Levin et al., 1976; Farrel et al., 1977). The conversion of proinhibitor (inactive eIF-2 kinase) to inhibitor (active eIF-2 kinase) can be promoted by a cAMP-dependent bovine heart protein kinase (Datta et al., 1977a). Furthermore hemin inhibits the activity of cAMP-dependent protein kinases from rabbit reticulocytes (Hirsch and Martelo, 1976). The involvement of cAMP-dependent protein kinase in translational control has been supported by experimental evidence (Sellers et al., 1974). Indeed, hemin prevents the binding of cAMP to the regulatory subunit of cAMP-dependent protein kinase from rabbit reticulocytes (Datta et al., 1977b). Therefore a general scheme

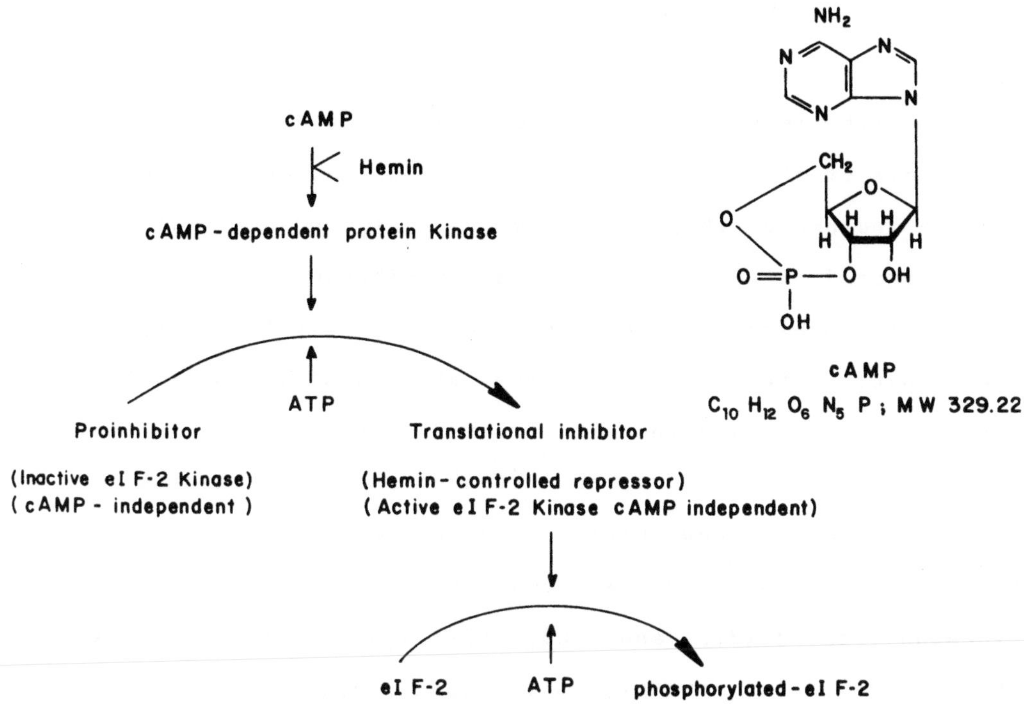

Fig. 5. Regulation of translation by cAMP (Datta et al., 1977b)

(Fig. 5) has been proposed to explain the translational control in reticulocytes (Datta et al., 1977b). A translational inhibitor, apparently acting as the hemin-controlled repressor does, has been reported in Friend leukemia cells (MW 214,000) (Pinphanichakarn, Kramer and Hardesty, 1977), ascites tumor cells (Clemens et al., 1976), and *Artemia salina* (Sierra et al., 1977), suggesting that such translational control is widely represented in eukaryotic cells.

The above results, therefore, suggest that cAMP (Fig. 5) might regulate protein synthesis in intact eukaryotic cells. Indeed, it was early shown that cAMP inhibits protein synthesis in liver slices (Pryor and Berthet, 1960) and cell-free systems from rat liver (Sellers et al., 1974). Correspondingly, there is evidence showing inhibition of cell growth when cAMP concentration in the cell is increased (Pastan, Johnson and Anderson, 1975; review).

Furthermore there are some results suggesting a possible role of cAMP in controlling phosphorylation of ribosomal protein S6 and hence ribosomal activity. Thus, phosphorylation of ribosomal protein S6 has been reported in ribosomal cell-free systems in the presence of cAMP (Loeb and Blat, 1970). Phosphorylation of protein S6 has been observed after addition of cAMP to rabbit reticulocytes (Cawthon et al., 1974) or rat liver (Gressner and Wool, 1974). Furthermore, an increased phosphorylation of protein S6 has been found under conditions where increased concentrations of cAMP have been observed, such as in the livers of glucagon-treated (Blat and Loeb, 1971) and diabetic (Gressner and Wool, 1976) rats. However, an increased phosphorylation of ribosomal protein S6 has also been observed in Krebs II ascites cells (Rankine and Leader, 1975) and baby hamster kidney fibroblasts under physiological conditions where the cellular concentration of cAMP is not elevated and the extent of the phosphorylation of protein S6 was also independent of alterations in the protein synthetic activity of the cells (Leader, Rankine and Coia, 1976). Indeed no significant differences between phosphorylated and diphosphorylated ribosomes have been observed in their protein synthetic activity (Eil and Wool, 1973). Furthermore, no phosphorylation of protein S6 was observed under conditions of ethionine-induced inhibition of protein synthesis and reversal of inhibition by treatment with adenine. However, cAMP concentration rose significantly in both ethionine-inhibited and adenine-reversed conditions (Treolar, Treolar and Kissilevsky, 1977).

2.4 Double-Stranded RNA and Interferon

Double-stranded RNA is formed in the replication of most RNA viruses and occurs as the genome of reovirus. Poliovirus double-stranded RNA is a strong inhibitor of protein synthesis by rabbit reticulocyte lysates in vitro (Ehrenfeld and Hunt, 1971). Furthermore, cellular protein synthesis is markedly inhibited by incubation of the cells in the presence of double stranded RNA (Cordell-Stewart and Taylor, 1973). These results suggested that viral double-stranded RNA might be responsible for virus interference with host cell protein synthesis. However, double-stranded RNA inhibits equally well in vitro mRNA from encephalomyocarditis virus and rabbit reticulocyte globin (Robertson and Matthews, 1973), and poliovirus double-stranded RNA inhibits cellular and viral mRNA translation to a similar extent in HeLa cell extracts (Celma and Ehrenfeld, 1974). Double-stranded RNA from reovirus or synthetic (poly I: poly C) inhibits translation of rabbit reticulocyte mRNA but does not affect protein synthesis by wheat embryo cell-free systems (Grill, Sun and Kandel, 1976).

Double-stranded RNA appears to prevent formation of the initiation complex on the 40S ribosomal subunit (Darnbrough, Hunt and Jackson, 1972) by inhibiting eIF-2; in rabbit reticulocyte systems this effect can be overcome by addition of an excess of eIF-2 (Kaempfer and Kaufman, 1973). Double-stranded RNA not only inhibits the initiation of protein synthesis but also induces the formation of interferon and can increase the effects of interferon in the infected cells; these effects have been briefly reviewed (Burke, 1977; review).

Interferons are glycoproteins synthesized by animal cells upon viral infection or treatment with interferon inducers such as double-stranded RNA. Interferon synthesis involves the induction of a cellular gene to form a new mRNA which is then translated and glycosylated. Interferons inhibit the multiplication of sensitive viruses apparently by preventing protein synthesis. Indeed extracts from interferon-treated cells are less active in the translation of both cellular and viral mRNA than extracts from control cells (Carter and Levy, 1967). This effect appears to be due to an impairment in peptide chain elongation caused by one inhibitor with nuclease activity loosely bound to ribosomes; since this inhibitor degrades the RNA, the inhibition can be partially overcome by the addition of tRNA (Friedman et al., 1972; Falcoff et al., 1973; Gupta, Sopori and Lengyel, 1973). Furthermore the methylation of the 5'-terminal cap of reovirus mRNA is impaired in interferon-treated cells (Sen et al., 1977).

Initiation of protein synthesis by cell-free extracts from interferon-treated cells is more sensitive to double-stranded RNA than that of similar extracts from control cells (Kerr, Brown and Ball, 1974). Thus in interferon-treated cells a low molecular weight inhibitor of initiation of translation is formed, requiring a pre-inhibitor, ATP,and double-stranded RNA; this low molecular weight inhibitor appears to be an oligonucleotide of unusual structure which is effective at subnanomolar concentrations, inhibiting protein synthesis by cell-free systems from mouse L-cells and rabbit reticulocytes (Roberts, Clemens and Kerr, 1976; Kerr, Brown and Hovanessian, 1977). However, the inhibitor is not formed when cells are treated with interferon and then exposed to double-stranded RNA before preparation of the cell-free system (Burke, 1977; review). The low molecular weight inhibitor has been formed in appropriate cell-free systems in response to double-stranded RNA and is characterized as having the structure pppA2'p5'A2'p5'A (Kerr and Brown, 1978).

In the presence of double-stranded RNA and ATP reovirus, mRNA is degraded faster in extracts from interferon-treated cells than in those from control cells; in these extracts double-stranded RNA promotes the phosphorylation by ATP of at least two proteins (Roberts, Clemens and Kerr, 1976; Lebleu et al., 1976; Ratner et al., 1977). The relationship between phosphorylation, the degradation of RNA (nuclease activity) induced by double-stranded RNA, and the low molecular weight inhibitor in extracts from interferon-treated cells has yet to be established.

3. Inhibitors of Initiator and mRNA Recognition [Steps (a) and (b)]

Within this group of inhibitors we include a number of triphenylmethane derivatives (the best known of which is aurintricarboxylic acid), some aminochromes (the best known of which is adrenochrome), polydextran sulphate, and polyvinyl sulphate (Fig. 6). All these compounds have a rather complex mode of action involving more than one interaction and inhibiting more than one step. Indeed all of them appear to interact with the ribosome and therefore prevent mRNA recognition [step (b)]. These compounds also inhibit recognition of the initiator [step (a)], at least in eukaryotic systems, by preventing

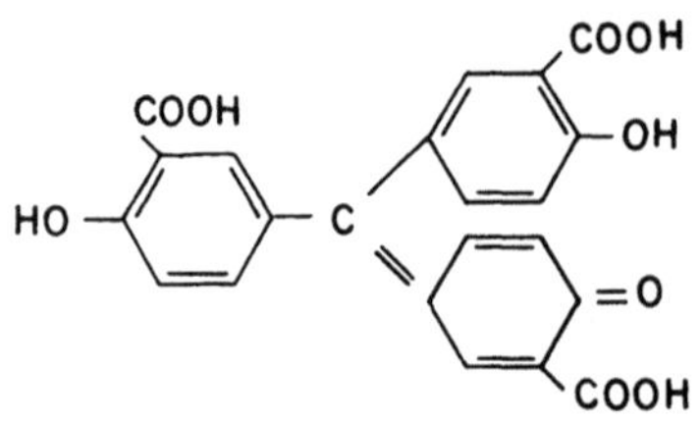

$C_{22}H_{14}O_9$; MW 422.35

Fig. 6. Chemical structure of aurintricarboxylic acid

the formation of the ternary complex eIF-2-Met-$tRNA_F$-GTP, in a reaction in which the ribosome is not involved.

3.1 Aurintricarboxylic Acid and Other Triphenylmethane Dyes

Aurintricarboxylic acid, and preferentially its ammonium salt (Fig. 6), is widely used as an inhibitor of initiation in cell-free systems since it interacts, even at low concentrations, with the small ribosome subunits (Huang and Grollman, 1972a) - both 30S and 40S - and blocks mRNA and synthetic polynucleotide binding to bacterial and eukaryotic ribosomes (Huang and Grollman, 1972a; Grollman and Stewart, 1968; Lebleu et al., 1970; Siegelman and Apirion, 1971; Heiberg, Olsnes and Pihl, 1971). Therefore at low concentrations aurintricarboxylic acid is a selective inhibitor of the initiation phase in bacterial (Huang and Grollman, 1973; Siegelman and Apirion, 1971; Stewart, Grollman and Huang, 1971), mammal (Huang and Grollman, 1972a; Lebleu et al., 1970; Heiberg, Olsnes and Pihl, 1971; Hoerz and McCarty, 1970; Stewart, Grollman and Huang, 1971), and plant (Marcus, Bewley and Weeks, 1970) systems, blocking step (b) in resolved assays with purified initiation factors (Fresno, Carrasco and Vázquez, 1976). However, the effect of aurintricarboxylic acid on this step is very small when tested in resolved systems with native 40S subunits (Fresno and Vázquez, 1978). Due to the permeability barrier aurintricarboxylic acid cannot be used, in most cases, in intact cells. Aurintricarboxylic acid also inhibits the formation of the eIF-3-Ac-Phe (Carrasco, Fernández-Puentes and Vâzquez, 1976) and eIF-2-Met-tRNA$_F$-GTP complexes (Stanley and Dettman, 1973; Fresno, Carrasco and Vázquez, 1976) and therefore appears to inhibit step (a) in resolved assays, at least in eukaryotic systems.

At low concentrations the effects of aurintricarboxylic acid in initiation in cell-free systems are rather specific, but at higher concentrations aurintricarboxylic acid blocks not only initiation but also elongation in bacterial and eukaryotic systems (Huang and Grollman, 1972a; Siegelman and Apirion, 1971) since it prevents polysome breakdown (Huang and Grollman, 1972a) and inhibits EF-G and ribosome-dependent GTPase and the function of EF-Ts (Weissbach and Brot, 1970).

Other triphenylmethane dyes such as pyrochatechol violet (Carrasco, Fernández-Puentes and Vázquez, 1976; Stanley and Dettman, 1973; Fresno, Carrasco and Vázquez, 1976; Lebleu et al., 1970) azure blue B, and to a lesser extent fuchsin acidic and fuchsin basic appear to have effects similar to those of aurintricarboxylic acid (Igarashi and Zmean, 1975). Indeed pyrochatechol violet blocks formation of the eIF-3-Ac-Phe-tRNA (Carrasco, Fernández-Puentes and Vázquez, 1976) and eIF-2-Met-tRNA$_F$-GTP (Stanley and Dettman, 1973; Fresno, Carrasco and Vázquez, 1976) complexes in eukaryotic cell-free systems and therefore prevents step (a) in resolved systems with derived subunits. Pyrochathecol violet also prevents ^{3}H-poly(U) binding to ribosomes (Huang and Grollman, 1973) and therefore blocks step (b) in initiation (Margulies and Kay, 1976).

3.2 Adrenochrome, Polydextran Sulphate, and Polyvinyl Sulphate

Adrenochrome (Fig. 7) and polydextran sulphate block formation of the eIF-3-Ac-Phe-tRNA (Carrasco, Fernández-Puentes and Vázquez, 1976) and eIF-2-Met-tRNA$_F$-GTP (Stanley and Dettman, 1973; Fresno, Carrasco and Vázquez, 1976) complexes in eukaryotic cell-free systems and therefore prevent the subsequent (a) step in resolved systems with derived subunits. Furthermore, polydextran sulphate (Miyazawa et al., 1967; Mathews and Korner, 1970) and polyvinyl sulphate (Shinozawa, Yahara and Imahori, 1968) by interacting with the small ribosomal subunit, prevent mRNA and polynucleotide binding to this subunit. These compounds are also inhibitors of step (b) in resolved assays with derived subunits (Fresno, Carrasco and Vázquez, 1976). As described above for aurintricarboxylic acid, and probably for the same reasons, they are also very poor inhibitors of initiation in cell-free systems using 40S native subunits (Fresno and Vázquez, 1978).

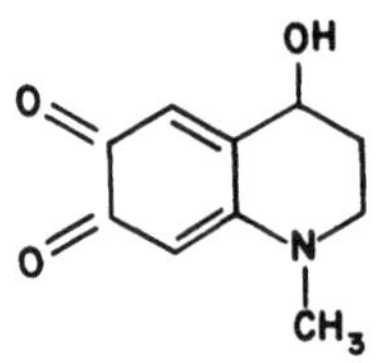

$C_9H_9O_3N$; MW 179.17

Fig. 7. Chemical structure of adrenochrome

4. Inhibitors of mRNA Recognition [Step (b)]

Edeine A_1 and 7-methylguanosine-5'-monophosphate are included within this group of inhibitors. Whereas edeine A_1 is active in bacterial and eukaryotic systems, 7'-methylguanosine-5'-monophosphate only inhibits translation by eukaryotic ribosomes of mRNA containing the $m^7G^{5'}$ppp moiety at their 5' end.

4.1 Edeine A_1

Edeine is an antibiotic complex produced by a strain of *Bacillus brevis* (Kurylo-Borowska, 1962, 1964). This complex preparation has broad spectrum activity, acting on gram-positive and gram-negative bacteria, some fungi and other eukaryotic cells as well as neoplastic cells (Kurylo-Borowska, 1962, 1964). This complex antibiotic contains four main components known as edeines A_1, A_2, B_1, and B_2 of which edeines A_1 and B_1 (Fig. 8) have antibiotic activity, whereas edeines A_2 and B_2 are rather inactive (Hettinger and Craig, 1970; Kurylo-Borowska and Szer, 1971). The basic constituents of edeines A_1 and A_2 are β-tyrosine, isoserine, diaminopropionic acid, 2,6-diamino-7-hydroxy-azelaic acid, glycine and spermidine, whereas spermidine is replaced by guanylspermidine

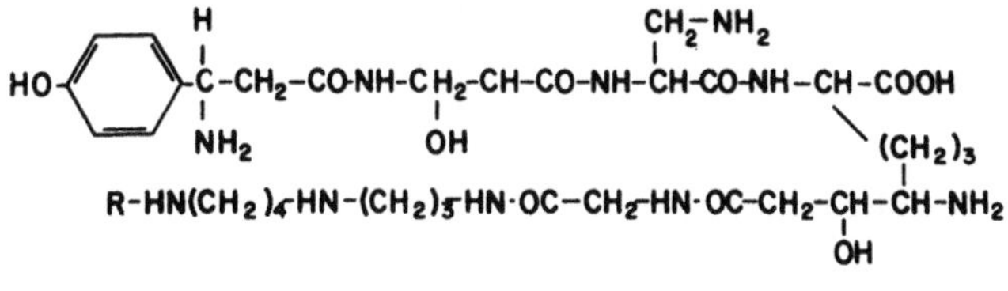

Edeine A_1 R = −H ($C_{33}H_{58}O_9N_{10}$; MW 738.85)

Edeine B_1 R = −C(=NH)NH_2 ($C_{34}H_{60}O_9N_{12}$; MW 780.89)

Fig. 8. Chemical structures of edeines A_1 and B_1

in edeines B_1 and B_2 (Hettinger and Craig, 1970). The α-NH_2 group of diaminopropionic acid is joined in the peptide link in the two active edeines (A_1 and B_1), whereas the β-NH_2 group of diaminopropionic acid is involved in this link in the two inactive edeines (A_2 and B_2) (Hettinger and Craig, 1970). In intact bacteria low concentrations of edeine reversibly inhibit DNA and enhance RNA synthesis whereas protein synthesis is unaffected (Kurylo-Borowska, 1962, 1964; Kurylo-Borowska and Szer, 1971). However, in cell-free systems edeine A_1 is a strong inhibitor of translation and structure activity studies have concluded that the action of the antibiotic is due to its unusual structure and not to its basicity (Kurylo-Borowska and Szer, 1971). Edeine A_1 interacts with both ribosomal subunits but has a preferential affinity for the small one (Vázquez, 1967a; Szer and Kurylo-Borowska, 1972). Edeine A_1 does not inhibit but rather enhances substrate recognition (either tRNA or Ac-Phe-tRNA or fMet-$tRNA_F$) by the donor site (Carrasco, Fernández-Puentes and Vázquez, 1976; Fresno, Carrasco and Vázquez, 1976; Obrig et al., 1976). Edeine A_1 does not affect peptide elongation in cell-free systems with the initiation complex preformed (Szer and Kurylo-Borowska, 1972; Obrig et al., 1976). Indeed the antibiotic is an inhibitor of initiation (Szer and Kurylo-Borowska, 1972; Obrig et al., 1976) which appears to prevent step (b) specifically (Fresno, Carrasco and Vázquez, 1976).

4.2 7-Methylguanosine-5'-Monophosphate

The compound $m^7G^{5'}p$ (7-methylguanosine-5'-monophosphate) specifically blocks translation of mRNA from tobacco mosaic virus, alphalpha mosaic virus (Roman et al., 1976; Hickey, Weber and Baglioni, 1976), brome mosaic virus (Shih, Dasgupta and Kaesberg, 1976), ovalbumin (Sharma, Hruby and Beezley, 1976) and different systems where mRNA is provided with the sequence $m^7G^{5'}$ ppp.... at the 5' end. This inhibition is due to a blockage at the initiation phase (Roman et al., 1976; Hickey, Weber and Baglioni, 1976) specifically preventing the eIF-4B-dependent interaction of mRNA with the 40S ribosome subunit (Shafritz et al., 1976). On the other hand $m^7G^{5'}p$ does not inhibit polypeptide elongation by polysomes (Wu, Cheung and Suhadolnik, 1977). A series of derivatives of the $m^7G^{5'}ppp$ structure have been comparatively tested for their inhibitory effects on translation by mRNA from tobacco mosaic virus, HeLa cells, and rabbit globin; it was observed that the 7-methyl and 5'-phosphate groups are essential for the inhibitory activity and that $m^7G^{5'}pp$ and $m^7G^{5'}ppp$ are better inhibitors than $m^7G^{5'}p$ (Hickey, Weber and Baglioni, 1976; Hickey et al., 1977), whereas a dinucleotide without the 7-methylated group is not inhibitory. It was proposed that the inhibitory activity of the $m^7G^{5'}p$ derivatives in protein synthesis occurs only in compounds where an interaction of a charge in the 7 position of guanosine with the negative charge of the 5'-phosphates is possible (Hickey et al., 1977). Studies on the effects of $m^7G^{5'}pppAm$, $m^7G^{5'}pppCm$, $m^7G^{5'}pppGm$, and $m^7G^{5'}pppUm$ on the synthesis of α-globin synthesis have shown that 50% inhibition was observed at six times lower concentrations than $m^7G^{5'}p$ (Suzuki, 1977).

Quantitative differences were observed in the inhibitory activities of m^7G derivatives in different systems. Thus $5'pm^7G$ and $m^7G^{5'}ppp^{5'}Np$ are strong inhibitors of translation of a number of stomatitis virus mRNAs tested in a wheat germ cell-free system for translation, but cause less than 20% inhibition in a reticulocyte system (Lodish and Rose, 1977).

5. Inhibitors of Subunit Joining [Step (c_1)]

Fluoride salts and 2-(4-methyl-2,6-dinitroanilino)-N-methyl-propionamide (MDMP) have been proposed as inhibitors of the joining of the 60S ribosomal subunit to the initiation complex formed on the 40S ribosomal subunit. No inhibitor of a similar initiation step in prokaryotic cells or cell-free systems has been described.

5.1 Fluoride Salts

Sodium fluoride inhibits translation, causing polysomal dissociation and accumulation of ribosomes and ribosomal subunits in mammalian cells and cell-free systems (Marks et al., 1965; Colombo, Vesco and Baglioni, 1968; Hogan, 1969; Bishop, 1968; Helgeland, 1976). The initiation complex with the 40S ribosomal subunits appears to be normally formed in the presence of sodium fluoride, but it has been proposed that this compound prevents the subsequent attachment of the 60S subunit to the complex (Lebleu et al., 1970; Hoerz and McCarty, 1970). This proposal has been strengthened by further experiments with potassium fluoride, which appears to have effects similar to those of sodium fluoride. Thus in the presence of potassium fluoride, chain elongation takes place but there is little initiation; polysome breakdown is observed and a complex is accumulated under these conditions which in the absence of the inhibitor can initiate polypeptide synthesis even in the presence of aurintricarboxylic acid (O'Rourke and Godchaux, 1975). The above evidence clearly suggests that sodium and potassium fluoride inhibit the joining of the 60S subunit to the initiation complex formed on the 40S subunit. Indeed in the presence of sodium fluoride unusual polysomes are formed in which one ribosome, presumably the last one to join the chain, has been replaced by a 40S subunit (Hoerz and McCarty, 1970). This explains the alteration of the ratio of free subunits to total ribosome population observed in animal cells treated with sodium fluoride (Vesco and Colombo, 1970). There is no evidence that fluoride at low concentrations inhibits any other step in translation, but this might be so at higher concentrations since chain elongation is partially inhibited and polysome breakdown is incomplete in the presence of potassium fluoride (O'Rourke and Godchaux, 1975).

There are no reports concerning the effects of sodium and potassium fluoride on intact bacteria or bacterial cell-free systems. Furthermore, no drug has been described as having in bacterial systems an effect similar to that of sodium fluoride in mammalian cells.

5.2 MDMP (2-(4-methyl-2,6-dinitroanilino)-N-methyl-propionamide)

A similar mechanism of action to that described above for fluoride has been proposed for the herbicide MDMP (Fig. 9). This proposal was based on detailed comparative studies with this compound and potassium fluoride (Weeks and Baxter, 1972). Only the D-MDMP stereoisomer blocks protein synthesis, whereas the L-MDMP has

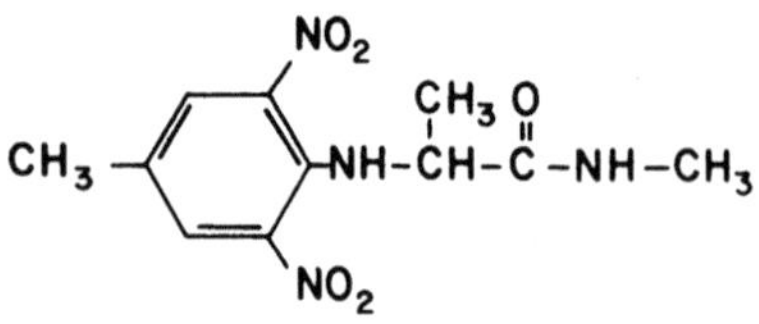

Fig. 9. MDMP

$C_{11} H_{14} O_5 N_4$; MW 282.25

no effect on this process (Weeks and Baxter, 1972) and neither competes with nor potentiates D-MDMP action (Baxter and McGowan, 1976). MDMP blocks translation in eukaryotic systems but not on 70S-type ribosomes. MDMP, like sodium and potassium fluoride, causes polysome breakdown in intact cells (Baxter and McGowan, 1976; Tscherne and Pestka, 1975) and this result is consistent with the mechanism of action proposed for the drug.

6. Inhibitors of the Positioning of the Initiator in the Donor Site [Step (c_2)]

The antibiotic pactamycin acts on the small ribosomal subunit and is active in both bacterial and eukaryotic systems.

6.1 Pactamycin

Pactamycin was isolated as an amorphous powder from culture filtrates of *Streptomyces pactum* (Bhuyan, Dietz and Smith, 1961) and the chemical structure was elucidated (Fig. 10) (Wiley and Mackellar, 1970). It is a very wide-spectrum antibiotic acting on gram-positive and gram-negative bacteria and also on different eukaryotic and tumoral cells (Bhuyan, Dietz and Smith, 1961).

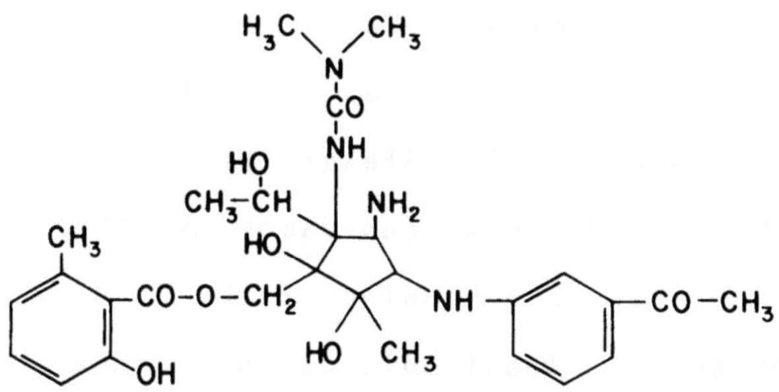

$C_{28}H_{38}O_8N_4$; MW 558.61

Fig. 10. Chemical structure of pactamycin

Pactamycin blocks protein synthesis by mammalian and bacterial systems (Colombo, Felicetti and Baglioni, 1966; Cundliffe and McQuillen, 1967) by interacting with the ribosomes (Felicetti, Colombo and Baglioni, 1966) on the smaller subunit (Cohen, Herner and Goldberg, 1969) as shown by binding experiments with the radioactive antibiotic (MacDonald and Goldberg, 1970; Stewart and Goldberg, 1973). There is one binding site for pactamycin per 30S subunit (Stewart and Goldberg, 1973). At higher concentrations pactamycin also binds to the larger ribosomal subunit (MacDonald and Goldberg, 1970). This lower affinity binding is maintained in polysomes bearing nascent peptides,but the higher affinity binding to the small subunit does not take place in these preparations (Stewart and Goldberg, 1973). It has been repeatedly observed that pactamycin does not inhibit polypeptide elongation by polysomes, since it causes polysome breakdown (Cundliffe and McQuillen, 1967; Ayuso and Goldberg, 1973; Scornik, 1974). There is however one conflicting report showing a similar inhibition by pactamycin on polypeptide elongation by E. coli endogenous polysomes and polypeptide synthesis directed by R17 using an E. coli cell-free system with initiation factors (Scornik, 1974). Studies on bacterial and mammalian systems have shown that pactamycin is an inhibitor of initiation (Cohen, Goldberg and Herner, 1969; Stewart-Blair, Yanowitz and Goldberg, 1971; Lodish, Houssman and Jacobsen, 1971) acting at some step taking place after the binding of the initiator Met-$tRNA_F$ to the 40S subunit (Baglioni, Jacobs-Lorena and Meade, 1972; Seal and Marcus, 1972; Kappen, Suzuki and Goldberg, 1973; Goldberg et al., 1973). Pactamycin does not inhibit but rather enhances Met-$tRNA_F$ binding to the 40S ribosome subunit, but interferes with the eIF-5-dependent joining of the 60S subunit (Fresno, Carrasco and Vázquez, 1976; Suzuki and Goldberg, 1974; Levin, Kyner and Acs, 1973), resulting in an accumulation of the initiation complex with the 40S subunit in resolved cell-free systems (Fresno, Carrasco and Vázquez, 1976; Kappen and Goldberg, 1976). However this effect may apparently be overcome in the presence of an excess of eIF-5 (Levin, Kyner and Acs, 1973; Kappen and Goldberg, 1976); a dipeptide is therefore accumulated in cell-free systems for polypeptide synthesis in the presence of pactamycin (Kappen and Goldberg, 1973; Cheung, Stewart and Gupta, 1973). This accumulation has been postulated as being due to the effect of the antibiotic in blocking the reuse and release of eIF-5 (Kappen and Goldberg, 1976).

7. Unclassified Inhibitors of Initiation

7.1 Streptomycin and Related Aminoglycoside Antibiotics

Streptomycin and most of its related aminoglycoside antibiotics are inhibitors of elongation of the polypeptide chain and therefore will be widely discussed in the pertinent chapter dealing with such inhibitors. However streptomycin, dihydrostreptomycin, and some other aminoglycosides also have an effect on initiation, or just after initiation, that will be discussed here.

Streptomycin has been shown to block initiation-dependent translation of viral and bacterial mRNA more effectively than amino acid incorporation on endogenous mRNA; it was therefore proposed that the antibiotic inhibits initiation (Modolell and Davis, 1968, 1969; Luzzatto, Apirion and Schlessinger, 1968, 1969). These results are consistent with the further report that a mutant with the Sm^D phenotype has an altered S12 protein and the main defect appears at the stage of step (c_2) (Lazar and Gros, 1973).

However, streptomycin does not block initiation itself, since 70S initiation complexes are formed in the presence of the antibiotic, but they are not stable and break down causing the release of f-Met-tRNA (Modolell and Davis, 1970a, 1970b; Lelong et al., 1971). It has been shown that streptomycin has an inhibitory effect on the dissociation activity of the initiation factor IF-3 on the ribosome (García-Patrone et al., 1971). Similarly ribosomes recovered from E. coli cells treated with streptomycin ("streptomycin-ribosomes") are impaired in their response to IF-3 (Wallace, Tai and Davis, 1973). It was therefore proposed that the antagonism of streptomycin to the dissociation effect of IF-3 on the ribosome causes the formation of abortive unstable initiation complexes and polysomes. Streptomycin-bound ribosomes are therefore released from initiation complexes and polysomes; these ribosomes are newly engaged in the cyclic abortive reinitiation that takes place (Wallace, Tai and Davis, 1973; Wallace and Davis, 1973).

Similarly to streptomycin the aminoglycoside antibiotics neomycin B and paromomycin also inhibit the dissociation activity of IF-3 on the ribosome (García-Patrone et al., 1971) and induce formation of unstable initiation complexes (Lando, Cousin and Garilhe, 1973); these results also suggest a close relationship of both effects of the antibiotics. However kanamycin inhibits the dissociation activity and has no effect on the stability of the initiation complex (García-Patrone et al., 1971; Lando, Cousin and Garilhe, 1973).

7.2 Pleuromutilin and Related Compounds

The antibacterial antibiotic pleuromutilin was isolated from different Basidiomycetes (*Pleurotus mutilis* and *Pleurotus passeckerianus*) and even from a species of *Drosophila* (*D. substrata*) (Glasby, 1976; Högenauer, 1978; reviews). It has an unusual chemical structure (Fig. 11) (Birch, Holzapfel and Rickards, 1966). Pleuromutilin is an antibacterial antibiotic active mainly on gram-positive bacteria and mycoplasms. Tiamulin, a compound closely related to pleuromutilin, is 10-50 times more active in some bacteria and mycoplasms (Högenauer, 1978; review).

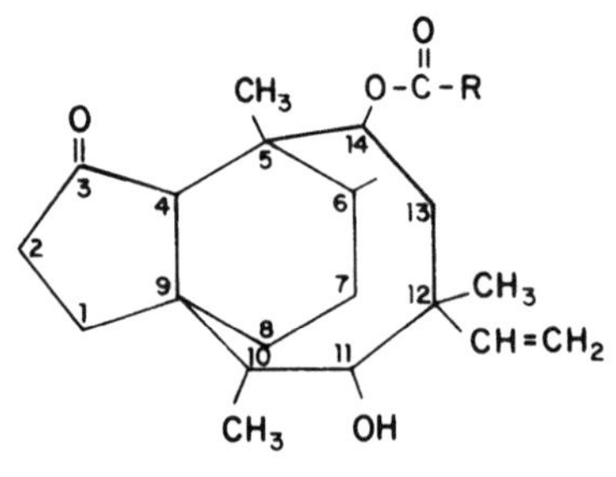

Pleuromutilin R = $-CH_2OH$ ($C_{22}H_{34}O_5$; MW 378.49)
Tiamulin R = $-CH_2-S-CH_2-CH_2-N(C_2H_5)_2$

Fig. 11. Chemical structures of pleuromutilin and tiamulin

The chemistry and mode of action of the derivatives of pleuromutilin have come under study in recent years (Egger and Reinshagen, 1976a, 1976b; Hodgin and Högenauer, 1974; Högenauer, 1975). The 14-deoxy-14-((2-diethylaminoethyl)-mercaptoacetoxy)-mutilin derivative (tiamulin) (Fig. 11) has been mainly used for studies on the mode of action of these antibiotics,since it is both very active and water-soluble (Hodgin and Högenauer, 1974). These antibiotics inhibit polypeptide synthesis by bacterial polysomes and poly (U)-directed polyphenylalanine synthesis (Hodgin and Högenauer, 1974). The pleuromutilin antibiotics appear to act as peptide bond formation inhibitors in model systems,since they inhibit the puromycin reaction (Hodgin and Högenauer, 1974). Furthermore binding of a labeled active derivative of pleuromutilin to the ribosome is inhibited by CCA, chloramphenicol, and puromycin but neither by lincomycin nor erythromycin (Högenauer, 1975). Pleuromutilin also inhibits EF-Tu-dependent binding of aminoacyl-tRNA to ribosomes in some systems but not in others (Hodgin and Högenauer, 1974).

Pleuromutilin does not affect formation of the initiation complex with the 30S ribosome subunit,although the initiation complex is readily dissociated in the presence of the antibiotic when the 50S subunit joins the complex (Högenauer, 1978; review). Therefore the antibiotic can inhibit initiation [possibly after step (c_2)] peptide bond formation and aminoacyl-tRNA binding. It is not known which of these effects is more relevant for the mode of action of the antibiotic, but the binding studies show that pleuromutilin acts on the 50S ribosomal subunit, preventing the interaction of the 3' terminal end of aminoacyl-tRNA with the acceptor site of the peptidyl transferase center (Högenauer, 1975).

7.3 5-Azacytidine

The base-analog 5-azacytidine (Fig. 12) primarily blocks translation in several biological systems, since it inhibits completely the synthesis of total protein and inducible enzymes while the overall rate of RNA synthesis remains nearly normal (Suhadolnik, 1970; review). The inhibitory effect of the compound in translation is located in HeLa cells at the initiation phase. Since the drug is incorporated into RNA and does not inhibit initiation of translation in resolved cell-free systems, it has been postulated that 5-azacytidine might be incorporated in a new RNA species, not yet described, somehow involved in the process of initiation (Reichman and Penman, 1973).

7.4 Thermorubin

Thermorubin is a complex antibacterial antibiotic that is produced by Thermoactinomyces antibioticus. The major component is thermorubin A ($C_{32}H_{24}O_{10}$; MW = 568.51) that is active against gram-positive and gram-negative bacteria. It inhibits initiation

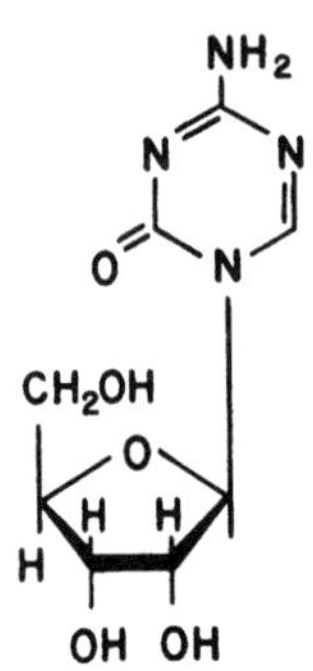

$C_8H_{12}O_5N_4$; MW 244.21

Fig. 12. Chemical structure of 5-azacytidine

of translation and has been shown to block polynucleotide-directed f-Met-$tRNA_F$ binding to bacterial ribosomes, but the specific step inhibited by the antibiotic has not been resolved (Pirali et al., 1974).

7.5 Inhibition of Translation After Viral Infection

Viral infection normally results in an inhibition of synthesis by the host cell (shut-off),whereas viral proteins are synthesized; no obvious explanation was given for these early observations. The shut-off effect has been observed after infection with a great variety of viruses including a number of bacteriophages (Condit, 1975) and different species of animal viruses (Carrasco, 1977; review). It does not require viral replication provided a high multiplicity of infection has been used (Bablanian, 1972).

The inhibition of protein synthesis after viral infection is exerted at some step not yet elucidated on the initiation level (Penman et al., 1963). None of the components required for translation appears to be affected in the infected cells as tested in cell-free systems (Lawrence and Thach, 1974; Abreu and Lucas-Lenard, 1976; Kaufman, Goldstein and Penman, 1976).

Double-stranded RNA was initially proposed as the possible specific inhibitor of cellular protein synthesis (Ehrenfeld and Hunt, 1971) but was later discarded since it inhibits cellular and viral mRNA translation in HeLa cell extracts to a similar extent (Celma and Ehrenfeld, 1974). It was also proposed that a still undefined inhibitor of mRNA binding to ribosomes could be generated after infection, differentially affecting cellular and viral mRNA translation (Nuss and Koch, 1976). It was further suggested that viral infection could interfere with some membrane function, thus altering the ionic concentration inside the cell, and that this ionic concentration might differentially affect viral and cellular protein synthesis (Carrasco, 1977). This hypothesis has been supported by experiments in cell-free systems showing that addition of sodium ions prevented translation of cellular mRNA and stimulated viral protein synthesis (Carrasco and Smith, 1976). It was further shown that it is likely that an alteration of the function of the membrane takes place after viral infection,since there is an impairment of the gradient concentration of monovalent ions inside and outside the cell (Carrasco and Smith, 1976; Egberts, Hackett and Traut, 1977). Moreover,whereas protein synthesis by 3T6 cells is resistant to the GTP analog GDPCP, since it cannot enter the eukaryotic cell, translation is prevented by the compound after the cells have been infected with picornaviruses, strongly suggesting that the membrane becomes leaky after viral infection and allows GDPCP to enter the cell (Carrasco, 1978). Similarly there are a number of protein synthesis inhibitors such as gougerotin that are only moderately active in 3T6 cells due to the permeability barrier, but become very active inhibitors of translation upon infection of the cells with picornaviruses (Contreras and Carrasco, 1978). The membrane becomes leaky to ions and small molecules not only in picornavirus-infected cells but also in a great variety of virus-cell systems, e.g., Semliki Forest virus infecting BHK cells and SV40 infecting CV1 cells (Carrasco, personal communication). A similar phenomenon also occurs in bacteria infected with bacteriophages (Condit, 1975).

The above experimental evidence therefore supports the proposal that the shut-off phenomenon, resulting in an inhibition of initiation of cellular protein synthesis, could be explained by the membrane leakiness model (Carrasco, 1977).

7.6 Inhibition of Protein Synthesis of Compounds Affecting the Function of the Plasma Membrane

At low concentrations, the ionophore antibiotic valinomycin ($C_{54}H_{90}O_{18}N_6$; MW = 1115.30), which is known to transport monovalent cations, causes a drastic inhibition of translation in cultured mammalian cells (Herzberg, Breitbart and Atlan, 1974; Breitbart et al., 1975). Furthermore the divalent cation ionophore A23187 is also a potent inhibitor of protein synthesis in C6 rat glioma cells, whereas the compound X537A (an ionophore antibiotic which transports divalent and monovalent cations) has only a slight inhibitory effect on translation (Bottenstein and de Vellis, 1976).

Phalloidin administration to rats induces a strong dissociation of liver polysomes and correspondingly blocks chain initiation. Similar effects were observed in hepatocyte cultures but phalloidin does not inhibit chain initiation in cell-free systems. It was proposed that the inhibitory effect of phalloidin in intact cells might be due to an effect of the drug on the membrane,causing a leakage of K^+ ions (Gravela and Poli, 1977).

It is not known whether the inhibitory effects of valinomycin, A23187, and phalloidin on protein synthesis are direct, or result from release of internal stores of cations (see above, inhibition of translation after viral infection). However it is surprising that the ionophore X537A, which is also very active in releasing monovalent and divalent cations, does not have such an inhibitory effect on translation. It would be interesting, then, to test similar effects on translation with other compounds which are known to affect the membrane.

7.7 Hypertonic Media

Hypertonicity resulting from an increased NaCl concentration in the growth medium of HeLa cells causes an inhibition of protein synthesis (Wengler and Wengler, 1972). Under these conditions polysome breakdown takes place and no effect has been detected on elongation or termination of the polypeptide chain; therefore it has been postulated that hypertonicity in the medium leads to an inhibition of peptide chain initiation (Saborio, Pong and Koch, 1974). Furthermore, in HeLa cells infected with picornaviruses, cellular protein synthesis was sensitive and viral protein synthesis was resistant to inhibition in hypertonic medium (Saborio, Pong and Koch, 1974; Tershak, 1976). Indeed it has been observed in different virus-cell systems that total cellular protein synthesis is much more sensitive to hypertonic media than viral protein synthesis. This has been shown with vesicular stomatitis virus (Nuss and Koch, 1976), proteins VP1, VP2 and VP3 (England, Howett and Tan, 1975), Semliki Forest virus (Clegg, 1975), and vaccinia virus (Oppermann and Koch, 1976). These results were initially interpreted by postulating that hypertonicity might release an inhibitor acting at the level of initiation and preferentially affecting cellular protein synthesis (Nuss and Koch, 1976). However a further and more plausible hypothesis suggests that the preferential inhibitory effect of hypertonicity might be due simply to a change in ionic concentration (Carrasco, 1977), since it has been shown that sodium ions inhibit cellular protein synthesis and enhance viral protein synthesis (Carrasco and Smith, 1976).

7.8 Temperature Shift-down

When exponentially growing cultures of bacteria, yeast, and higher cells are cooled to below 8°C, initiation of protein synthesis is blocked, while elongation of the polypeptide chains continues until completion (Friedman, Lu and Rich, 1969; McCormick and Penman, 1969; Barbacid and Vázquez, 1975).

7.9 3-Methyleneoxindole

The compound 3-methyleneoxindole was initially proposed as a selective inhibitor of viral protein synthesis (Tuli, 1974), but it has been shown to inhibit synthesis of cellular and viral proteins to a similar extent (Abreu and Lucas-Lenard, 1977). The compound has been proposed as an inhibitor of initiation since it does not affect poly(U)-directed polyphenylalanine synthesis but blocks translation of natural mRNA; furthermore there is a similar pattern of inhibition of translation with 3-methyleneoxindole and aurintricarboxylic acid (Abreu and Lucas-Lenard, 1977).

7.10 Cl^- Ions

In mammalian systems Cl^- ions at concentrations higher than those found in the cytoplasm (70-80 mM) have been reported to inhibit translation by preventing the attachment of mRNA to the ribosomes and therefore blocking the initiation phase (Weber et al., 1977).

7.11 Nitrofurans (Nitrofurantoin)

The nitrofurans are a large group of synthetic antibacterial compounds extensively used in medical treatment and as food additives. The best known compound of this group is nitrofurantoin (Fig. 13). Recent experimental evidence suggests that nitrofurans inhibit specifically the initiation of translation of only certain mRNAs; this inhibition appears to concern only the expression of inducible genes (Herrlich and Schweiger, 1976; Wagner et al., 1977).

7.12 2'(3'),5'-ADP

At adequate concentrations 2'(3'),5'-ADP has litle or no effect on polypeptide elongation by cell-free systems from Ehrlich ascites tumor cells, but inhibits initiation. The

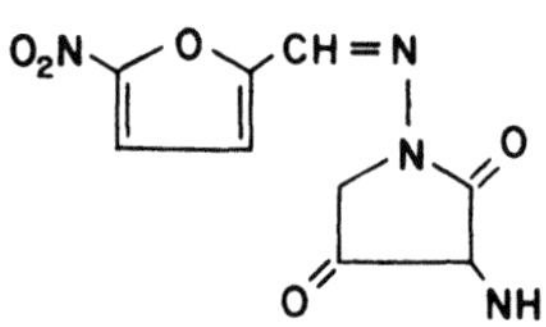

$C_8H_6O_5N_4$; MW 238.16

Fig. 13. Chemical structure of nitrofurantoin

specific step blocked by 2'(3'),5'-ADP has not been elucidated, but it certainly inhibits prior to formation of the first peptide bond (Buchwald et al., 1977).

7.13 Other Antibiotics and Compounds Proposed as Inhibitors of Initiation

D-amphetamine sulphate administered to rats decreases brain protein synthesis and causes polysome breakdown (Moskowitz et al., 1975). Neuroleptics such as haloperidol and chlorpromazine can effectively reverse this polysome breakdown (Fig. 14) (Widelitz, Coryell and Avadhani, 1977). Polypeptide elongation by brain or liver polysomes is not inhibited by the drug. On the other hand D-amphetamine and p-chloroamphetamine block amino acid incorporation in a crude initiation-dependent system in wheat germ. (^{3}H)amphetamine binds to wheat germ ribosomes but this binding depends on the presence of mRNA, suggesting that the drug block of initiation takes place at a step of initiation after mRNA binding (Baliga et al., 1976). These results were confirmed in a rabbit reticulocyte cell-free system, further showing that the order of potency of several para-substituted amphetamines was pOH<pF≃amphetamine<pCl<Br. Moreover, at their inhibitory concentrations all these compounds inhibit aminoacylation of tRNA, suggesting that this is the step

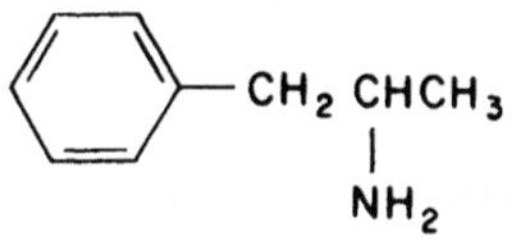

$C_9H_{13}N$; MW 135.20 Fig. 14. Chemical structure of D-amphetamine

specifically inhibited by the amphetamines (Nowak and Munro, 1977).

Antibiotics of the thiostrepton group (thiostrepton, siomycin, sporangiomycin, and thiopeptin) have been reported to inhibit initiation in bacterial systems. Since the effects of these antibiotics have been studied extensively in the elongation process, we will discuss their action on initiation together with their effects on aminoacyl-tRNA binding.

Lincomycin, spiramycin, and the streptogramin A group of antibiotics (active on bacterial systems), verrucarin A, toxin T-2, deacetoxyscirpenol, fusarenon X, bruceantin, and harringtonine (active on eukaryotic cells) have been proposed as initiation inhibitors on the grounds that they allow ribosome run-off from polysomes. This observation is based on the fact that these antibiotics do not interact with polysomes and therefore allow polysome breakdown. However, they do interact with bacterial or eukaryotic ribosomes and are strong inhibitors of peptide bond formation in the first few rounds of elongation prior to polysome formation. We will therefore discuss these inhibitors within the group of compounds blocking step (e).

The antibiotics enomycin and phenomycin have been proposed as inhibitors of initiation. We will discuss their mode of action within the group of elongation inhibitors, since they certainly inhibit step (d).

Viomycin has been shown to inhibit the initiation process in E. coli cell-free systems, but we will discuss its mode of action within the elongation inhibitors since it also inhibits translocation (Liou and Tanaka, 1976; Modolell and Vázquez, 1977).

Cycloheximide, pederine, abrin, ricin, colicin E_3, and cloacin DF_{13} have been reported to inhibit initiation. However, since these drugs also inhibit elongation and the step affected by the antibiotics in initiation has not been resolved, we will discuss their effects within the group of elongation inhibitors.

Chapter 3
Elongation

1. Introduction

Polypeptide elongation takes place by repeated cycles composed of three sequential steps: EF-Tu (or EF-1)-dependent aminoacyl-tRNA binding [step (d)], ribosome-catalyzed peptide bond formation [step (e)] and EF-G (or EF-2)-dependent translocation [step (f)]. Besides the initiation complex formed in step (c_2), two molecules of GTP, the specific aminoacyl-tRNA and the elongation factors (EF-Ts, EF-Tu and EF-G in bacteria; EF-1 and EF-2 in eukaryotes) are required in the first elongation cycle.

There is ample experimental evidence supporting a single entry site in intact cells for GTP- and EF-Tu (or EF-1)-dependent binding of aminoacyl-tRNA on the A-site of the ribosome. Non-enzymic binding of aminoacyl-tRNA to the A-site can also be studied in cell-free systems in the absence of EF-Tu and EF-1.

Step (e) is catalyzed by the peptidyl transferase center, which is an integral part of the larger ribosomal subunit (50S in bacterial and 60S in eukaryotic ribosomes).

Translocation [step (f)] requires one of the elongation factors (EF-G in bacteria and EF-2 in eukaryotes) and involves movement of the peptidyl-tRNA from the A- to the P-site, coupled to the movement of one triplet of the mRNA, on the ribosome. Hydrolysis of one molecule of GTP is required, probably for the release of EF-G (or EF-2). Spontaneous non-enzymic translocation takes place slowly in cell-free systems in the absence of elongation factors and GTP. This non-enzymic translocation is considerably accelerated in eukaryotic systems in the presence of high concentrations of K^+ or NH_4^+ .

The two elongation factors complexed with GTP appear to interact with a common or overlapping area of the larger ribosomal subunit and therefore do not interact simultaneously but only alternately with the ribosome. Thus the interaction of one of the elongation factors takes place after the release of the other.

2. Compounds Interfering with Aminoacyl-tRNA Recognition [Step (d)]

Within this important category of inhibitors we can distinguish at least two groups of compounds. One of these groups includes the inhibitors of EF-Tu (or EF-1)-dependent binding of aminoacyl-tRNA and the other one clusters a number of compounds that prevent protein synthesis by favoring the erroneous interaction of aminoacyl-tRNA in the acceptor site of the ribosome. This phenomenon, which is now widely known as misreading, is also called miscoding by some workers.

2.1 Inhibitors of Aminoacyl-tRNA Binding

Some of the inhibitors in this group act on the small ribosomal subunit (the tetracyclines),whereas TPCK inactivates EF-Tu, fusidic acid freezes EF-G (or EF-2) on the ribosome and most of the others act on the larger ribosomal subunit (antibiotics of the thiostrepton group, abrin, ricin, PAP, and alpha sarcin). Since the EF-Tu (or EF-1)-aminoacyl-tRNA-GTP complex is formed prior to the binding reaction, in this step of binding some of the inhibitors appear to interfere specifically at the ribosome level with the interaction or the function of the elongation factor EF-Tu or EF-1, whereas others appear to prevent the correct positioning of the aminoacyl-tRNA.

Within this particular class of inhibitors we include a number of glycoproteins and proteins (ricin, abrin, alpha sarcin, PAP, crotins, curcins, enomycin, and phenomycin) which catalytically inactivate the 60S ribosomal subunit of eukaryotic ribosomes.

2.1.1 The Tetracycline Antibiotics

The best known members of this group of closely related antibiotics are tetracycline, oxytetracycline (terramycin), chlortetracycline (aureomycin), doxycycline, minocycline, and 5a, 6-anhydrotetracycline (Fig. 15), which appear to have a

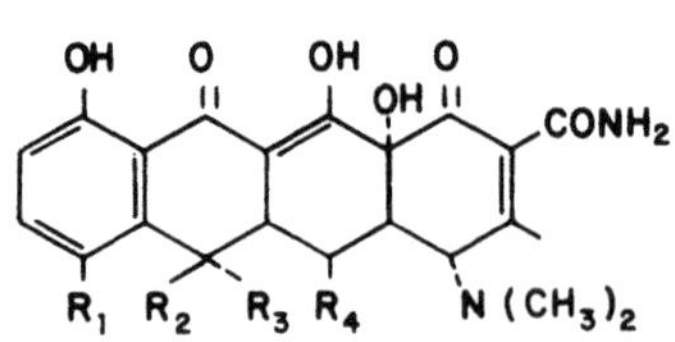

Chlortetracycline	: R_1 = -Cl;	R_2 = -OH;	R_3 = -CH_3;	R_4 = -H	($C_{22}H_{23}O_8N_2Cl$; MW 478.88)
Demethylchlortetracycline	: R_1 = -Cl;	R_2 = -OH;	R_3 = -H ;	R_4 = -H	($C_{21}H_{21}O_8N_2Cl$; MW 464.05)
Oxytetracycline	: R_1 = -H ;	R_2 = -OH;	R_3 = -CH_3;	R_4 = -OH	($C_{22}H_{24}O_9N_2$; MW 460.44)
Tetracycline	: R_1 = -H ;	R_2 = -OH;	R_3 = -CH_3;	R_4 = -H	($C_{22}H_{24}O_8N_2$; MW 444.43)
Doxycycline	: R_1 = -H ;	R_2 = -H ;	R_3 = -CH_3;	R_4 = OH	($C_{22}H_{24}O_8N_2$; MW 444.43)

Fig. 15. Chemical structure of the tetracycline antibiotics

similar mode of action at the ribosome level. All these antibiotics have a broad inhibitory spectrum, being active in cell-free systems of prokaryotes and eukaryotes. However, these antibiotics are preferentially accumulated in intact bacteria and are therefore widely used against infections by gram-positive and gram-negative bacteria.

Oxytetracycline was initially obtained from Streptomyces rimosus (Finlay et al., 1950) and chlortetracycline was elaborated from Streptomyces aureofaciens cultures (Duggar, 1948). Tetracycline was initially obtained by hydrogenation of chlortetracycline and later from cultures of several Streptomyces species, whereas doxycycline and minocycline are semisynthetic antibiotics which have the advantage in clinical use of being absorbed better by mammalian tissues in some cases. The chemical structures of the tetracycline antibiotics were early established (Stephens et al., 1954) and their absolute configuration has been elucidated.

The tetracycline antibiotics block protein synthesis in bacterial and mammalian systems (Rendi and Ochoa, 1961; Franklin, 1963). Since these antibiotics were found to prevent non-enzymic binding to bacterial ribosomes (Hierowski, 1965; Suárez and Nathans, 1965) and to their 30S subunits (Suzuka, Kaji and Kaji, 1966; Pestka and Nirenberg, 1966; Vázquez and Monro, 1967), it was assumed that their main effect was on the small ribosomal subunit. However, the EF-Tu-dependent binding of aminoacyl-tRNA to 70S ribosomes is inhibited by much lower concentrations of the tetracyclines than the non-enzymic binding to 30S subunits (Lucas-Lenard and Haenni, 1968; Ravel et al., 1968; Gordon, 1969; Skoultchi et al., 1970; Modolell et al., 1971). Thus enzymic binding of aminoacyl-tRNA is the specific step inhibited by low concentrations of the tetracyclines, although the EF-Tu-dependent GTP hydrolysis is not affected by the antibiotics (Lucas-Lenard and Haenni, 1968; Ravel et al., 1968; Gordon, 1969; Skoultchi et al., 1970; Modolell et al., 1971).

Early studies showed that tetracycline antibiotics have many interaction sites with both ribosomal subunits (Connamacher and Mandel, 1965; Day, 1966; Maxwell, 1966) but it is not yet clear which interaction is more important for their inhibitory effect on protein synthesis. Thus the strong interaction of the tetracyclines reported to occur with the 50S subunit might be more relevant to their mode of action than the weak and strong interactions of the antibiotics with the 30S subunit (Fey, Reiss and Kersten, 1973). This is still a matter of controversy since a specific interaction with the 30S ribosomal subunit has been observed at low concentrations of tetracyclines (Connamacher and Mandel, 1968). Indeed it has been reported that the strong type of binding of the tetracyclines to the ribosome specifically prevents the interaction of aminoacyl-tRNA (Streltsov et al., 1974). Furthermore, recent equilibrium data using fluorescence methods suggest that there is only one strong binding site for the drug on the ribosome and that this binding site is sufficient to inhibit one ribosome; the data suggest that Mg^{2+} chelation is not important in the mechanism of action of tetracycline (Tritton, 1977). This proposal contradicts a previous report suggesting that the tetracyclines bind to ribosomes only unspecifically, forming chelates with magnesium ions and therefore affecting the structure and function of the ribosome (White and Cantor, 1971).

Besides the blocking of protein synthesis, there are a number of properties of inhibitory effects which might differ among tetracycline derivatives. Thus doxycycline and minocycline are much more lipophylic than the classical tetracyclines (tetracycline, oxytetracycline,and aureomycin) (Barza et al., 1975). This might explain both why man absorbs them better in some cases and their activity in some _Staphylococcus aureus_ resistant to more hydrophylic tetracyclines. Furthermore, 5a, 6-anhydrotetracycline not only blocks protein synthesis, but also inhibits synthesis of RNA, DNA and pyrimidine triphosphates (Silverman and Atherly, 1978). All tetracyclines having antibacterial activity inhibit the ribosome-dependent systems synthesizing ppGpp (Pedersen, Lund and Kjeldgaard, 1973; Cochran and Byrne, 1974; Beres and Lucas-Lenard, 1975) but some of them (doxycycline and 5a, 6-anhydrotetracycline) are weak inhibitors of ppGpp synthesis in a ribosome-free stringent factor assay (Silverman and Atherly, 1977). At rather high concentrations the tetracycline antibiotics also inhibit a number of cellular processes such as bacterial oxidation and respiration and the activities of a number of enzymes (Laskin, 1967; review).

2.1.2 Antibiotics of the Thiostrepton Group and Multhiomycin (Nosiheptide)

Thiostrepton (synonyms, bryamycin and thiactin) (MW ≈ 1650) (Bodanszky, Dutcher and Williams, 1963), siomycin (MW ≈ 1712), thiopeptin (MW ≈ 1750), sporangiomycin (MW ≈ 1800) and to a lesser extent nosiheptide are related sulphur containing antibiotics. The chemical structures of thiostrepton, siomycin and nosiheptide are already known (Fig. 16) (Anderson, Hodgkin and Viswamitra, 1970; Prange et al., 1977; Tori et al., 1976). Nosiheptide is produced by Streptomyces actuosus (Prange et al., 1977), siomycin is produced by Streptomyces syoyaensis (Nishimura et al., 1961), thiostrepton by an unclassified species of Streptomyces (Pagano et al., 1955-1956), sporangiomycin by Planomonospora parontospora (Thiemann et al., 1968), and thiopeptin by Streptomyces tateyamensis (Miyairi et al., 1970). The antibiotics are active only on gram-positive bacteria, but in bacterial cell-free systems they act on ribosomes from gram-positives and gram-negatives.

Antibiotics of the thiostrepton group appear to have a similar mode of action. Early genetic studies suggested that the

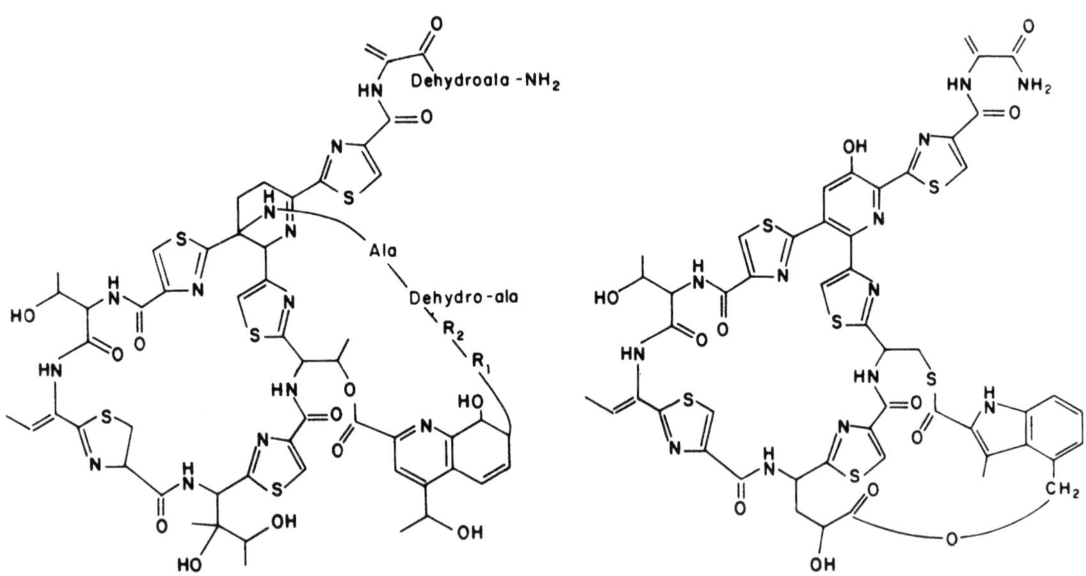

Thiostrepton (R_1-R_2: Isoleucine-alanine)
($C_{72}H_{85}N_{19}O_{18}S_5$; MW 1664,85

Siomycin (R_1-R_2: Valine-dehydroalanine)

Nosiheptide ($C_{51}H_{43}N_{13}O_{12}S_6$; MW 1222,33)

Fig. 16. Chemical structures of thiostrepton, siomycin, and nosiheptide

action of these antibiotics on Bacillus subtilis might be located on the ribosome (Dubnau et al., 1967). This was later confirmed (Goldthwaite and Smith, 1972; Bazzicalupo et al., 1975), showing that the resistance is associated with the 50S subunit (Liou et al., 1973) and that a ribosomal protein is altered in mutants resistant to the antibiotic (Pestka et al., 1976). Furthermore ribosomes from thiostrepton-sensitive strains revertant from thiostrepton-resistant strains bind ^{35}S-thiostrepton in a similar way to ribosomes from the parental sensitive strain (Vince et al., 1976). The interaction of the thiostrepton antibiotics with the ribosomes was confirmed by biochemical studies showing that they bind to the 50S subunit (Weisblum and Demohn, 1970; Tanaka, K., et al., 1970; Modolell, Vázquez and Monro, 1971; Pirali et al., 1972; Highland, Howard and Gordon, 1975). Quantitative studies on binding of ^{35}S-labeled thiostrepton have shown that the interaction (one molecule of the antibiotic per ribosome) is specific for the 50S subunit, is relatively independent of pH, NH_4^+, and Mg^{2+} concentrations but is inhibited by pre-bound EF-G (Highland, Howard and Gordon, 1975). Thiostrepton enhances EF-Tu- or EF-G-dependent GTP hydrolysis by 50S-derived cores lacking the ribosomal proteins L7 and L12 (Ballesta and Vázquez, 1972a), showing that these proteins are not directly involved in the binding of the antibiotic (Brot et al., 1972; Sopori and Lengyel, 1972). Protein L11 has been reported to be required for the interaction of thiostrepton with the ribosome, since the loss in antibiotic binding was correlated with the removal of this protein, and the ability of some L11-deprived 50S cores to bind thiostrepton was restored by the addition of protein L11 and prevented by the presence of antibodies prepared against this protein (Highland et al., 1975). However,thiostrepton interacts with some ribosome cores specifically deprived of protein L11, inhibiting their function; it is therefore obvious that this protein is not an absolute requirement for thiostrepton binding (Bernabeu, Vázquez and Ballesta, 1976; Brot et al., 1973; Schrier and Moller, 1975) although in its absence the affinity of the ribosome for thiostrepton is certainly lower (Bernabeu, Vázquez and Ballesta, 1976). Furthermore, alteration of protein L5 has been observed in a thiopeptin-resistant mutant (Liou and Tanaka, 1975), whereas a mutant strain of B.subtilis possesses an altered form of a ribosomal protein that appears to be the counterpart of protein L11 of E. coli ribosomes (Pestka et al., 1976; Wienen et al., 1977). On the other hand Streptomyces azureus, the microorganism producer of thiostrepton, protects itself against the antibiotic by methylating the 23S RNA of its 50S ribosomal subunit (Cundliffe, 1978).

The thiostrepton antibiotics inhibit translocation (Tanaka, K., et al., 1970; Pestka, 1970b) and EF-G-dependent GTP hydrolysis (Modolell, Vázquez and Monro, 1971; Bodley, Lin and Highland, 1970; Kinoshita, Liou and Tanaka, 1971; Watanabe, 1972) in model systems by preventing the interaction of EF-G with the ribosome (Modolell, Vázquez and Monro, 1971; Cundliffe and Dixon, 1975); they were therefore initially postulated as specific inhibitors of the translocation step. However, it was observed in further studies that the thiostrepton antibiotics inhibit both enzymic (Modolell et al., 1971; Kinoshita, Liou and Tanaka, 1971) and non-enzymic binding of aminoacyl-tRNA to the ribosomal A-site in cell-free systems (Modolell et al., 1971; Watanabe, 1972). Unlike the tetracyclines, the thiostrepton antibiotics block the GTP hydrolysis coupled to EF-T-dependent aminoacyl-tRNA binding (Modolell et al., 1971). The EF-Tu-dependent GTP hydrolysis uncoupled from aminoacyl-tRNA binding is insensitive to thiostrepton (Ballesta and Vázquez, 1972b), indicating that these antibiotics block the A-site on the 50S subunit thus preventing the attachment of EF-G and aminoacyl-tRNA. Comparative studies have shown that the thiostrepton antibiotics inhibit the EF-G-dependent reaction and EF-T-dependent binding of aminoacyl-tRNA in cell-free systems (Modolell et al., 1971; Watanabe, 1972; Weissbach et al., 1972) to a similar extent. Since thiostrepton does not inhibit translocation in intact bacteria (Cannon and Burns, 1971), the main target reaction of the antibiotic in vivo is aminoacyl-tRNA binding (Modolell et al., 1971; Cundliffe, 1971).

^{35}S-thiostrepton binding to bacterial ribosomes is strongly inhibited by siomycin, sporangiomycin, and thiostrepton, further demonstrating the similarity of the antibiotics of the thiostrepton group in their mode of action (Cundliffe and Dixon, 1975). Furthermore the Streptomyces producer of thiostrepton, as well as its ribosomes in cell-free systems, are resistant not only to thiostrepton but also to siomycin and sporangiomycin (Dixon, Beven and Cundliffe, 1975).

It has been suggested that there is a common site on the ribosome involved in the interaction of GTP complexed with protein factors involved in translation (Modolell and Vázquez, 1975, review). Indeed in resolved cell-free systems the thiostrepton antibiotics inhibit not only EF-Tu- and EF-G-dependent GTP hydrolysis but also the GTPase activity catalyzed by IF-2 (Grunberg-Manago, Dondon and Graffe, 1972). As a consequence of this action the thiostrepton antibiotics inhibit initiation in a complex manner. Thus thiostrepton does not inhibit f-Met-$tRNA_F$ binding to the ribosome or IF-2- or ribosome-dependent uncoupled GTP hydrolysis when IF-2 functions stoichiometrically; rather, it prevents f-Met-$tRNA_F$ binding and coupled GTP hydrolysis when IF-2 functions catalytically (Mazumder, 1973; Lockwood et al., 1974). Therefore the antibiotic does not directly inhibit GTP hydrolysis in initiation, but blocks the action of IF-1 which is required for the recycling of IF-2 (Sarkar, Stringer and Maitra, 1974). Indeed thiostrepton prevents ^{14}C-IF-2 interaction with the 70S-initiation complex and inhibition of initiation by the antibiotic is strengthened in the presence of IF-1 (Naaktgeboren, Vermaas and Voorma, 1975).

Multhiomycin is a sulphur containing antibiotic that is obtained from Streptomyces antibioticus (Tanaka, T., et al., 1970). Identity of multhiomycin with nosiheptide has been demonstrated recently (Endo and Yonehara, 1978). Multhiomycin interacts with the bacterial ribosome and inhibits EF-Tu-dependent aminoacyl-tRNA binding and GTP hydrolysis (Tanaka, Sakaguchi and Yonehara, 1970; Tanaka, Sakaguchi and Yonehara, 1971). All the experimental evidence, although scarce, suggests that multhiomycin can be considered as a member of the thiostrepton group although the molecule is smaller.

2.1.3 Micrococcins and Thiocillins

There are a number of antibacterial antibiotics that appear to be closely related to thiostrepton in their chemical structure and mode of action including multhiomycin, the micrococcins, and the thiocillins.

Micrococcin is produced by a bacterial species resembling Micrococcus varians (Su, 1948) and by Bacillus pumilus (Abraham et al., 1956); it is a mixture of micrococcins P_1 (M_1) (MW ≈ 1119) and P_2 (M_2) (Shoji et al., 1976). The thiocillins (I, II and III) (thiocillin I, MW ≈ 1292; thiocillin II, MW ≈ 1392) are antibacterial antibiotics, produced by different species of Bacillus which are closely related to each other and also to the micrococcins (Shoji et al., 1976).

The locus concerned with resistance to micrococcin was mapped early on and shown to be in a region of the Bacillus subtilis genome containing a number of ribosomal proteins, suggesting that the antibiotic might be an inhibitor of protein synthesis (Dubnau et al., 1967). This was confirmed by showing that ribosomes from a micrococcin-resistant mutant are resistant to the antibiotic (Goldthwaite and Smith, 1972), the resistance is associated with the 50S subunit (Smith, Weiss and Piestka, 1976) and a ribosomal protein is altered in mutants resistant to the antibiotic (Wienen et al., 1977). The loci for micrococcin and thiostrepton are not precisely the same since ribosomes from the micrococcin-resistant mutant are resistant to micrococcin and sensitive to thiostrepton (Pestka et al., 1976) and alterations in different proteins are involved in micrococcin and thiostrepton resistance (Wienen et al., 1977).However,these loci appear to be closely related since ribosomes from the Streptomyces strain producing thiostrepton are resistant to thiostrepton and micrococcin (Dixon, Beven and Cundliffe, 1975) and the thiostrepton-resistant mutants of Bacillus subtilis are resistant to thiostrepton and micrococcin (Goldthwaite and Smith, 1972). Micrococcin was proposed as an inhibitor of translocation (Pestka and Brot, 1971; Otaka and

Kaji, 1974), as previously suggested for thiostrepton (see above), and also shown to prevent EF-Tu-dependent binding of aminoacyl-tRNA (Otaka and Kaji, 1974). Unlike thiostrepton, micrococcin does not inhibit EF-G-dependent uncoupled GTP hydrolysis (Pestka and Brot, 1971); like thiostrepton it appears to prevent substrate interaction with the A-site on the 50S subunit. However micrococcin is a rather poor inhibitor of thiostrepton binding (Cundliffe and Dixon, 1975) as might be expected from the genetic studies (Wienen et al., 1977), since the loci for micrococcin and thiostrepton are not the same.

2.1.4 Fusidic Acid

Fusidic acid (Fig. 17) is a steroid antibiotic produced by *Fusidium coccineum* which is active on gram-positive bacteria and mycobacteria (Godfredsen et al., 1962). Fusidic acid is active in both prokaryotic and eukaryotic cell-free systems. Its chemical structure is already established (Godfredsen et al., 1965).

Fusidic acid can prevent EF-G- and EF-2-dependent translocation and uncoupled GTP hydrolysis in bacterial and

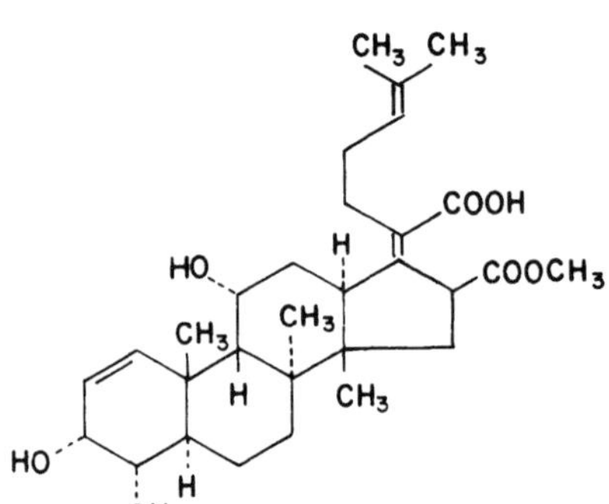

$C_{31}H_{48}O_6$; MW 516.69

Fig. 17. Chemical structure of fusidic acid

mammalian systems (Tanaka, Kinoshita and Masukawa, 1968; Tanaka et al., 1969; Malkin and Lipmann, 1969; Haenni and Lucas-Lenard, 1968; Pestka, 1968). Fusidic acid probably acts on EF-G since EF-G is altered in bacterial mutants resistant to the antibiotic (Kinoshita, Kawano and Tanaka, 1968; Tocchini-Valentini and Mattoccia, 1968; Bernardi and Leder, 1970). EF-G was also suggested as the target for the action of the antibiotic by preliminary binding experiments using tritiated fusidic acid (Okura, Kinoshita and Tanaka, 1970; Okura, Kinoshita and Tanaka, 1971), although it was reported in further work with (^{3}H)-24,25 dihydrofusidic acid that ribosomes and GTP or GDP are required for the interaction of the antibiotic with EF-G (Willie et al., 1975). Whereas the 24, 25 dihydro derivative is as active as fusidic acid, saturation of the double bond between C_{17} and C_{20} drastically reduces the activity of the antibiotic (Willie et al., 1975; Bodley and Godfredsen, 1972). In the presence of ribosomes (or their larger subunits), GTP, and either EF-G or EF-2, fusidic acid forms a stable fusidic acid-GDP-ribosome-EF-G (or EF-2) complex, which accounts for the inhibitory effect of the antibiotic on translation (Bodley et al., 1969; Bodley et al., 1970; Bodley and Lin, 1970; San Millán, Vázquez and Modolell, 1975). GTP and GDP are equally active in the formation of this complex (Willie et al., 1975; Brot, Spears and Weissbach, 1971). Nevertheless, fusidic acid allows one round of GTP hydrolysis and translocation; this, then, is not the step inhibited by the antibiotic in intact cells or protein synthesizing systems. Thus fusidic acid-inhibited polysomes, formed on synthetic or viral RNA or on endogenous mRNA, have their nascent peptides in the puromycin-reactive position (Celma, Vázquez and Modolell, 1972;

Modolell and Davis, 1970b; Pestka and Hintikka, 1971; Cundliffe, 1972). Furthermore, formation of the GDP-ribosome-fusidic acid-EF-G complex (Cabrer, Vâzquez and Modolell, 1972; Miller, 1972; Richman and Bodley, 1972; Richter, 1972) prevents aminoacyl-tRNA binding. Similar inhibition has been obtained when the GDP-ribosome-fusidic acid-EF-2 complex is preformed (Carrasco and Vâzquez, 1973a; Otaka and Kaji, 1973; Richter, 1972). Formation of these fusidic acid complexes prevents the interaction of EF-Tu (or EF-1) with the ribosome since EF-Tu and EF-G (or EF-1 and EF-2) appear to have a common or overlapping interaction site on the larger ribosome subunit (Cabrer, Vâzquez and Modolell, 1972; Miller, 1972; Richman and Bodley, 1972; Richter, 1972). Therefore although fusidic acid interacts with the elongation factor involved in translocation (EF-G or EF-2), the antibiotic does not inhibit translocation in synthesizing systems but rather aminoacyl-tRNA binding. In cell-free systems fusidic acid, by forming the GDP-ribosome-fusidic acid-EF-G (or EF-2) complex, can sequester all the EF-G or EF-2 available (Cabrer, Vâzquez and Modolell, 1972; Miller, 1972; Richman and Bodley, 1972; Richter, 1972). Thus the effect of the antibiotic in cell-free systems is decreased or abolished by saturating concentrations of EF-G (Kaji, Igarashi and Ishitsuka, 1969; Modolell, Cabrer and Vâzquez, 1973b; Burns, Cannon and Cundliffe, 1974) or EF-2 (Carrasco and Vâzquez, 1973a), whereas it is enhanced increasing the concentration of free ribosomes (Carrasco and Vâzquez, 1973a). The reported inhibition of translocation by fusidic acid is only observed in model systems of this step in which EF-G (or EF-2) is sequestered in the fusidic acid complex and is therefore not available to catalyze the translocation reaction (Carrasco and Vâzquez, 1973a; Burns,

Cannon and Cundliffe, 1974; Modolell, Cabrer and Vázquez, 1973b). Although with less affinity, fusidic acid can also form the complex with EF-G, GTP and polysomes with peptidyl-tRNA bound to either the A- or the P-site (San Millán, Vázquez and Modolell, 1975); therefore aminoacyl-tRNA binding is also inhibited by fusidic acid in translating polysomes as a consequence of the formation of the complex (San Millán, Vázquez and Modolell, 1975).

2.1.5 Ricin and Abrin

Ricin and abrin are very similar in their chemical structure and mode of action. A number of reviews have recently been published covering extensively the toxic properties, chemical structures, functions, and modes of action of ricin, abrin, and the related lectins *Ricinus communis* agglutinin and *Abrus precatorius* agglutinin (Olsnes and Pihl, 1977) as well as the mode of action of different glycoprotein and protein toxins acting on eukaryotes (Vázquez, 1978b) (Tables 6 and 7).

Ricin is a glycoprotein of 65,000 - 66,000 molecular weight isolated from seeds of *Ricinus communis*. It has also been equated with ricin D, ricin RCA_{60}, ricin II, RCA_{II}, and ricin toxin (Olsnes and Pihl, 1977; Nicolson, Blaustein and Etzler, 1974), although some workers do not agree that these are one and the same compound (Lugnier and Dirheimer, 1973). Recent crystallographic studies propose a MW of 65,000 for ricin (Villafranca and Robertus, 1977).

Abrin is a glycoprotein of 65,000 molecular weight isolated from the seeds of *Abrus precatorius* (Olsnes and Pihl, 1973a). It forms tetrameric structures detectable by X-ray diffraction studies (McPherson and Rich, 1973).

Table 6. Glycoproteins and proteins that specifically block translation

Compound	Producer organism	Chemical structure	
Colicin E3	*Escherichia coli* (CA 38)	Protein	(MW=60,000)
Cloacin DF13	*Enterobacter cloacae* (Clo DF13)	Protein	(MW=56,000)
Diphtheria toxin	*Corynebacterium diphtheriae* (βtox+)	Protein	(MW=62,000)
PA toxin	*Pseudomonas aeruginosa* (PA-103)	Protein	(MW=54,000)
Ricin	Seeds of *Ricinus communis*	Glycoprotein	(MW=66,000)
Abrin	Seeds of *Abrus precatorius*	Glycoprotein	(MW=65,000)
Crotin II	Seeds of *Croton tiglium*	Protein? or glycoprotein?	
Curcin II	Seeds of *Jatropha cursas*	Protein? or glycoprotein?	
PAP	Leaves of *Phytolacca americana*	Protein	(MW=31,000)
Alpha sarcin	*Aspergillus giganteus*	Protein	(MW=16,800)
Mitogillin	*Aspergillus restrictus*	Protein	(MW=16,200)
Restrictocin	*Aspergillus restrictus*	Protein	(MW=16,300)
Enomycin	*Streptomyces mauvecolor*	Protein	(MW=11,000)
Phenomycin	*Streptomyces fervens* var. *phenomyceticus*	Protein	(MW=10,000)

Neither ricin nor abrin causes appreciable cell agglutination although there is a strong effect with both *Ricinus communis* agglutinin (also known as RCA_{120} and ricin I) and *Abrus precatorius* agglutinin which have molecular weights of 120,000 and 134,000 respectively and are rich in carbohydrates

Table 6 (continued)

Spectra activity	Specific action
Prokaryotic systems	Splitting 49 nucleotides of 3' terminus of 16S ribosomal RNA
Prokaryotic systems	Splitting 48-49 nucleotides of 3' terminus of 16S ribosomal RNA
Eukaryotic systems	ADP-ribosylation of elongation factor EF-2
Eukaryotic systems	ADP-ribosylation of elongation factor EF-2
Eukaryotic systems	Catalytic inactivation of the 60S ribosomal subunit
Eukaryotic systems	Catalytic inactivation of the 60S ribosomal subunit
Eukaryotic systems	Catalytic inactivation of the 60S ribosomal subunit
Eukaryotic systems	Catalytic inactivation of the 60S ribosomal subunit
Eukaryotic systems	Catalytic inactivation of the 60S ribosomal subunit
Eukaryotic systems	Catalytic inactivation of the larger RNA of the 60S ribosomal RNA
Eukaryotic systems	Catalytic inactivation of the larger RNA of the 60S ribosomal RNA
Eukaryotic systems	Catalytic inactivation of the larger RNA of the 60S ribosomal RNA
Eukaryotic systems	Catalytic inactivation of the 60S ribosomal subunit
Eukaryotic systems	Catalytic inactivation of the 60S ribosomal subunit

(Olsnes and Pihl, 1972a; Nicolson, Blaustein and Etzler, 1974; Lugnier et al., 1974; Olsnes et al., 1975a).

The high toxicity of ricin and abrin on eukaryotic cells is a result of their action in specifically blocking protein synthesis. The toxins have no effect on RNA synthesis

Table 7. The effects of glycoprotein and protein toxins acting on the eukaryotic ribosome on model reactions

	EF-1-dependent		EF-2-dependent			
	Binding of aminoacyl-tRNA	GTP hydrolysis	Formation of the EF-2-GTP-ribosome complex	Uncoupled		Non-enzymic
				GTP hydrolysis	Translocation	translocation
Ricin A-chain	Inhibition	Inhibition	Inhibition	No effect or inhibition	No effect or stimulation	No effect
Abrin A-chain	Inhibition	Inhibition	Inhibition	No effect or inhibition	No effect or stimulation	No effect
Alpha sarcin	Inhibition	Inhibition	Inhibition	Inhibition	No effect	---
PAP	Inhibition	Inhibition	Inhibition	No effect or stimulation	No effect or stimulation	---
Crotin II	Inhibition	Inhibition	Inhibition	No effect	No effect or inhibition	---
Modeccin	Inhibition	---	Inhibition	---	---	---
Enomycin	Inhibition	---	No effect	Inhibition	No effect	---

Data taken from different references as indicated in the text.

and their inhibitory action on DNA synthesis is a secondary consequence of their specific blockade of protein synthesis (Lin et al., 1970a, 1971). Accordingly, both ricin and abrin are potent antitumor drugs in mice bearing Ehrlich ascites cells (Lin et al., 1970b; Olsnes, Refsnes and Pihl, 1974). Specific inhibition of protein synthesis by ricin and abrin has also been confirmed with HeLa cells (Olsnes and Pihl, 1972a), plasmacytoma cells (Ko and Kaji, 1975), mouse myeloma cells (Onokazi, Hayatsu and Ukita, 1975), and normal- and virus-transformed fibroblasts (Nicolson, Lacorbiere and Hunter, 1975). In all these cases there is a lag of 30-60 min before protein synthesis is halted by the toxins.

Mutants of Chinese hamster ovary cells and baby hamster kidney fibroblasts, both resistant to ricin, have been isolated. Although some of these mutants bind ricin only poorly, others do so as efficiently as normal sensitive cells (Gottlieb, Skinner and Kornfeld, 1974; Meager, Ungkitchanuki and Hughes, 1976). A number of cell line mutants resistant to ricin and abrin become sensitive and bind higher amounts of ricin after neuraminidase treatment, showing that these mutations do not affect ribosome sensitivity (Rosen and Hughes, 1977; Sandvig, Olsnes and Pihl, 1978). A further inhibitory effect of ricin is related to the interaction of the B-subunit with the cell membrane; thus the drug has been reported to decrease the level of cyclic AMP normally enhanced by preincubation of Yoshida ascites sarcoma with prostaglandin E_1 (Nomoto, Satoh and Ukita, 1975).

Both ricin and abrin contain two polypeptide chains joined by a S-S bond (Fig. 49) (Funatsu, 1972; Nicolson, Blaustein and Etzler, 1974; Olsnes, Refsnes and Pihl, 1974), but important differences exist between the chemical structures of the two drugs. Thus, abrin A chain does not contain carbohydrate whereas ricin A chain contains approximately 2.4% carbohydrate mainly in the form of four residues of mannose (Olsnes, Refsnes and Pihl, 1974; Olsnes et al., 1975b). However, both abrin B chain and ricin A chain contain some 7% carbohydrate which is in the form of 10-11 residues of mannose and 1-2 residues of acetyl glucosamine (Olsnes, Refsnes and Pihl, 1974; Olsnes et al., 1975b). There are also some differences in the amino acid composition of the two drugs (Funatsu, 1972; Nicolson, Blaustein and Etzler, 1974; Olsnes et al., 1975b), which in addition differ immunologically (Ehrlich, 1957; Olsnes, Pappenheimer and Meren, 1974) since the isolated A- and B-chains of each toxin do not crossreact. The heterologous chains abrin A/ ricin B and ricin A/abrin B can easily be formed and are as toxic as ricin and abrin (Olsnes, Pappenheimer and Meren, 1974). However, whereas anti-abrin or anti-ricin serum neutralizes only its homologous toxin, neither antitoxin can distinguish between the hybrid forms although antibodies specifically directed against determinants on A- or B-chains readily distinguish between the two hybrids (Olsnes, Pappenheimer and Meren, 1974).

The treatment of ricin and abrin with 2-mercaptoethanol reduces the S-S bond joining their A- and B-chains and the subunits separate. As a result, treatment of the toxins with 2-mercapto-ethanol abolishes both their toxicity to animals and their inhibitory effect on protein synthesis in intact eukaryotic cells (Olsnes and Pihl, 1972b), although if the 2-mercaptoethanol is removed the disulphide bonds reform and toxicity is restored (Olsnes, Refsnes and Pihl, 1974). In contrast separation of the A- and B-chains by treatment with 2-mercaptoethanol results in a considerable increase of the toxic activity in cell-free systems (Olsnes and Pihl, 1972b). Apparently the free A-chain is a very active inhibitor in cell-free systems, whereas the B-chain is essentially ineffective (Olsnes and Pihl, 1973a, 1973b). Indeed all the evidence suggests that A-chains have to be separated from B-chains to cause inhibition of protein synthesis since, even at high concentrations, the intact toxins do not block translation in cell-free systems in the absence of reducing agents. This provides an explanation for the slow kinetics of inhibition shown by the toxins in cell-free systems (Olsnes and Pihl, 1972a, 1972c; Grollman et al., 1974). In addition, however, it has been reported that a small tryptic peptide from ricin (MW = 12,500) can block translation in rabbit reticulocyte cell-free systems (Lugnier et al., 1974).

The B-chains of ricin and abrin, but not the A-chains, bind to erythrocytes (Olsnes and Pihl, 1973a, 1973b). Furthermore the B-chains are retained by Sepharose 4B columns and can be eluted with buffers containing galactose (Olsnes and Pihl, 1973b; Olsnes, Refsnes and Pihl, 1974). Indeed, ricin and abrin bind to both free galactose and lactose with a single binding site for the sugars on their B-chains (Olsnes and Pihl, 1977). It has been shown that (^{125}I)-labeled ricin and abrin bind reversibly to both HeLa cells and erythrocytes and again the interaction is prevented or reversed by galactose or lactose (Olsnes and Pihl, 1977), these sugars drastically reducing the inhibitory effect of the toxins (Onozaki et al., 1972; Olsnes, Refsnes and Pihl, 1974; Grollman et al., 1974). It is clear from these results that the B-subunit is required for the attachment of the toxin to the cells in order to facilitate the entry of the A-subunit. This subunit can be freed by reduction of the toxin inside the cell, and is then able to block protein synthesis (Refsnes, Olsnes and Pihl, 1974).

The A-subunits of both ricin and abrin inhibit translation by inactivating ribosomes (Olsnes and Pihl, 1972b; Olsnes, 1972; Olsnes, Heiberg and Pihl, 1973; Montanaro, Sperti and Stirpe, 1973; Onozaki, Hayatsu and Ukita, 1975; Benson et al., 1975). Thus (^{3}H)-labeled ricin A subunits bind to both 80S ribosomes and their corresponding 60S subunits but associate with neither 40S subunits nor bacterial ribosomes (Hedblom, Cawley and Houston, 1976). There are conflicting reports concerning the effects of the toxins on different types of ribosomes. Thus ricin has been reported to block endogenous protein synthesis in both *Neurospora crassa* and yeast mitochondria (Lugnier, Kuntzel and Dirheimer, 1976) but was found to be ineffective on poly(U)-directed polyphenylalanine synthesis by ribosomes from rat liver mitochondria (Greco et al., 1974). Furthermore some workers have observed that ribosomes from all eukaryotic cells tested, including plants, green algae, and yeast, are sensitive to the toxins, while bacterial ribosomes are not affected (Olsnes, Heiberg and Pihl, 1973; Greco et al., 1974). It was reported more recently, though, that ricin A is not able to inhibit polyphenylalanine synthesis by wheat germ ribosomes at 5000 times the concentration that is effective in blocking synthesis in a similar system by rat liver ribosomes; both wheat germ and rat liver ribosomes bind the toxin to a similar extent, however (Cawley et al., 1977).

Inhibition of 60S ribosomal subunits by ricin and abrin is catalytic rather than stoichiometric, since a single molecule of the toxin can inactivate a large number of ribosomes

(Olsnes and Pihl, 1972a, 1972b, 1972c; Sperti et al., 1973; Carrasco, Fernández-Puentes and Vázquez, 1975). The A-subunits of the toxins can, in fact, inactivate ribosomes in buffer solutions at a rate of 1,500 ribosomes per minute per toxin A-subunit (Olsnes et al., 1975a) and this result clearly shows that the A-subunit acts enzymically. The nature of the enzymic activity has not been elucidated to date since neither modifications of the ribosomal proteins (Nicolson, Lacorbiere and Hunter, 1975; Lugnier et al., 1976) nor of ribosomal RNA (Mitchell et al., 1976) by the toxins or their A-subunits have been demonstrated.

There are contradictory reports concerning which events in translation are inhibited by the toxins. Since polysome breakdown was observed in extracts from both rat liver (Lin et al., 1972) and HeLa cells (Grollman et al., 1974) after a certain period of toxin treatment, it was postulated that ricin inhibits initiation of protein synthesis (Grollman et al., 1974). However, ricin and abrin have been reported to stabilize polysomes in both rabbit reticulocyte lysates (Olsnes and Pihl, 1972a; Nicolson, Blaustein and Etzler, 1974) and mouse myeloma cells (Onozaki, Hayatsu and Ukita, 1975),suggesting that in these systems the toxins prevent elongation of the polypeptide chain. Ricin and abrin do not in fact affect peptide bond formation but inhibit both the EF-1-dependent binding of aminoacyl-tRNA (Carrasco, Fernández-Puentes and Vázquez, 1975) and GTP hydrolysis (Benson et al., 1975; Sperti et al., 1975) without affecting the non-enzymic binding of aminoacyl-tRNA (Carrasco, Fernández-Puentes and Vázquez, 1975). These results clearly suggest that the toxins inactivate one site on the 60S ribosomal subunit involved in the interaction or function of EF-1. This site appears to be identical to or to overlap the ribosomal binding site for EF-2, since the toxins also inhibit formation of an EF-2-GTP-ribosome complex with washed ribosomes (Carrasco, Fernández-Puentes and Vázquez, 1975; Montanaro et al., 1975; Nolan, Grasmuk and Drews, 1976). In certain cases the toxins were reported to inhibit partially ribosome- and EF-2-dependent uncoupled GTP hydrolysis (Montanaro, Sperti and Stirpe, 1973; Benson et al., 1975; Sperti et al., 1975), but other workers have observed that this reaction is stimulated by the toxins (Sperti et al., 1976; Fernández-Puentes and Vázquez, 1977). Nevertheles inhibition of EF-2 binding by the toxins was confirmed by replacing GTP with either GDP or GMPPCP (Carrasco, Fernández-Puentes and Vázquez, 1975); prior attachment of EF-2 and aminoacyl-tRNA to ribosomes protects these particles from the inactivation that normally results from treatment with the toxins (Mitchell et al., 1976; Fernández-Puentes, Carrasco and Vázquez, 1976). In addition the inhibition of translation observed with toxin-treated ribosomes is obscured or abolished by increasing concentrations of EF-1, EF-2, or ribosomes (Fernández-Puentes, Carrasco and Vázquez, 1976). It appears, therefore, that the damage caused by ricin and abrin on the 60S ribosomal subunit does not totally inactivate the particle but causes a considerable decrease in the affinity between ribosomes and both EF-1-GTP and EF-2-GTP. Thus the toxins have no significant effect when an excess of either EF-1 or ribosomes is used (Montanaro et al., 1975; Sperti et al., 1975; Fernández-Puentes, Carrasco and Vázquez, 1976) although EF-1-dependent GTP hydrolysis and aminoacyl-tRNA binding are very sensitive to ricin and abrin when concentrations of EF-1 and ribosomes are limiting (Benson et al.,

1975; Carrasco, Fernández-Puentes and Vázquez, 1975; Sperti et al., 1975; Fernández-Puentes, Carrasco and Vázquez, 1976). Ricin and abrin inhibit neither EF-2-dependent translocation (Montanaro, Sperti and Stirpe, 1973; Stirpe et al., 1976; Fernández-Puentes, Carrasco and Vázquez, 1976) nor non-enzymic translocation induced by high concentration of K^+ (L. Carrasco, personal communication), since the substrate bound to the acceptor site protects the ribosome from inactivation by the toxin (Fernández-Puentes, Carrasco and Vázquez, 1976). This explains why a large number of ribosomes derived from polysomes of treated cells escape inactivation by the toxin (Fodstad and Olsnes, 1977).

A complex between ribosomal 5S RNA and a single ribosomal protein, sedimenting at 8S and derived from 60S ribosomal subunits, has low GTPase activity which is inhibited by both ricin and abrin A-subunits. This complex might form the specific ribosomal target site for the toxin (Benson et al., 1975; Benson, 1977).

The results summarized above indicate that both ricin and abrin block aminoacyl-tRNA binding by preventing the attachment of EF-1-GTP-aminoacyl-tRNA complexes to ribosomes, thereby inhibiting elongation of polypeptide chains (Fig. 18) although they cannot prevent translocation. Common or overlapping sites located on 60S ribosomal subunits are involved in the attachment of GTP, itself bound to certain initiation, elongation, and termination factors. Inactivation of this ribosomal site by ricin and abrin might also result in inhibition of the initiation and termination processes (Fig. 18). Such effects remain to be demonstrated in resolved assay systems, but preliminary experiments in a crude system suggest that abrin A chain might prevent the

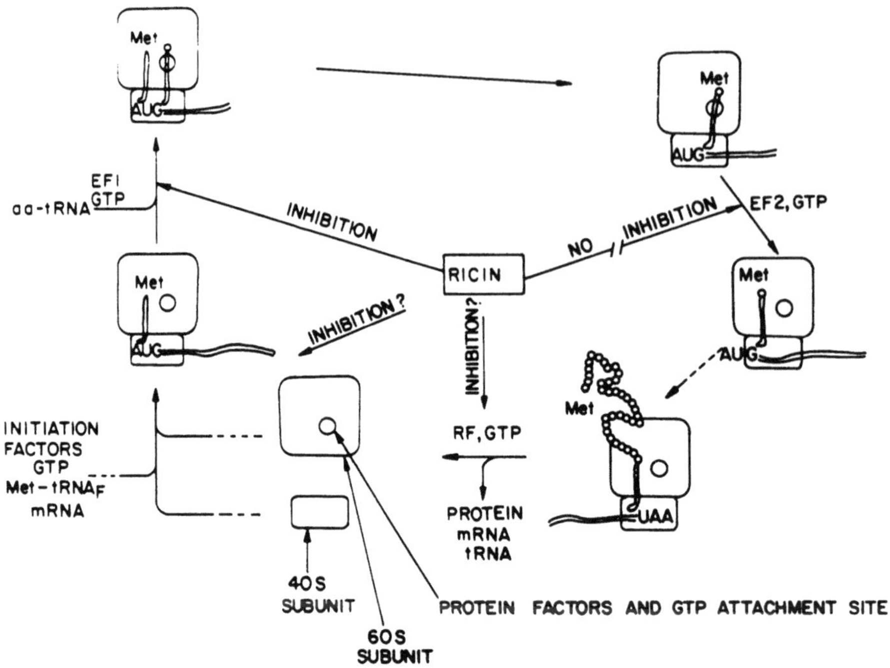

Fig. 18. Ribosome cycle showing possible sites of inhibition by ricin

joining of the 60S subunit [step (c_1)] to the initiation complex f-Met-tRNA·80S (Skorve et al., 1977). This result appears to be at variance with another report showing no activity of ricin A-chain on wheat germ ribosomes (Cawley et al., 1977); there may be a different spectrum of activity between ricin A chain and abrin A chain.

2.1.6 PAP, Alpha Sarcin, Restrictocin, Mitogillin, Crotins, Curcins, Enomycin, Phenomycin, and Modeccin

PAP, alpha sarcin, restrictocin, mitogillin, enomycin, phenomycin, modeccin, and the individual crotins and curcins (Tables 6 and 7) are similar to both ricin A chain and abrin A chain in that their inhibitory activity is rather low in intact cells, probably because of cellular permeability barriers. The mode of action of these compounds has been reviewed recently (Vázquez, 1978b). However, they are very active inhibitors in cell-free systems and much less than one molecule of each toxin per ribosome is required to inhibit translation, suggesting that, as is the case for the A-subunits of both ricin and abrin, all these proteins inactivate ribosomes enzymically. Indeed in comparative studies with rabbit reticulocyte and yeast systems it has been shown that abrin, PAP, alpha sarcin, and enomycin have in common an inhibitory effect on EF-1-dependent binding of aminoacyl-tRNA and do not inhibit EF-2-dependent translocation of peptidyl-tRNA (Fernández-Puentes and Vázquez, 1977).

The crotins are proteins extracted from the seeds of Croton tiglium, which are very active in blocking formation of EF-2-GTP-ribosome complexes whilst enhancing EF-2-dependent uncoupled hydrolysis. The purified crotins, mainly crotin II, inhibit EF-1-dependent binding of aminoacyl-tRNA. It was proposed that croton II inhibited the translocation step in protein synthesis, but the drug has a very poor effect on this reaction only (Sperti et al., 1976). The interaction of crotins with ribosomes may not be prevented by peptidyl-tRNA bound to the acceptor site, a situation which would contrast with that observed for both ricin and abrin A chain.

The curcins are proteins extracted from the seeds of Jatropha curcas and the compounds appear to have a mode of action similar to that of the crotins. It is interesting that the individual curcins are much less active than the curcin complex (Sperti et al., 1976).

PAP is a basic protein of MW = 31,000 according to X-ray diffraction studies (Robertus, Monzing and Irvin, 1977). It is extracted from the leaves of Phytolacca americana (Wyatt and Shepherd, 1969) and blocks EF-1-dependent reactions involving both GTPase activity and binding of aminoacyl-tRNA to ribosomes (Obrig, Irvin and Hardesty, 1973; Suits and Irvin, 1976; Fernández-Puentes and Vázquez, 1977). PAP also inhibits formation of the EF-2-GTP-ribosome complex but it has no effect on, and even enhances, EF-2-dependent GTP hydrolysis (Obrig, Irvin and Hardesty, 1973; Fernández-Puentes and Vázquez, 1977). Cell-free systems from Artemia salina are more sensitive to PAP than those from reticulocyte lysates, and this differential sensitivity appears to be related to differences in the EF-2 factor in each system (Suits and Irvin, 1976). (^{14}C) Phe-tRNA binding to the acceptor site protects the ribosomes from inactivation by PAP (Dallal and Irvin, 1978). PAP has an antiviral effect that probably results from its activity in preventing protein synthesis (Tomlinson et al., 1974).

Alpha sarcin is a basic protein (MW = 16,000) which contains the unusual amino acid sarcinine. It is produced in the fermentation broth of Aspergillus giganteus. Alpha sarcin is an active antitumor agent and is cytotoxic though hardly active against E. coli (Olson and Goerner, 1965; Olson et al., 1965). Restrictocin and mitogillin are also antitumor proteins which contain sarcinine and are very similar to alpha sarcin immunologically since they are cross reactive (Conde et al., 1978). Alpha sarcin blocks protein synthesis in extracts from wheat germ, yeast, rat liver, rabbit reticulocytes and Ehrlich ascites cells by inhibiting both EF-1-dependent GTP hydrolysis and binding of aminoacyl-tRNA to ribosomes, but it has no effect on the non-enzymic binding of aminoacyl-tRNA to ribosomes (J.E. Davies, pers. communication; Fernândez-Puentes and Vâzquez, 1977). Thus alpha sarcin probably blocks translation by preventing the functional interaction of EF-1-GTP-aminoacyl-tRNA complexes with ribosomes. Interaction of the toxin with ribosomes in the pretranslocated state does not appear to take place, as is also the case for abrin and ricin; alpha sarcin does not, therefore, inhibit translocation. Catalytic inactivation of the 60S ribosomal subunit by alpha sarcin results from enzymic cleavage, by the protein, of 320 nucleotides from the 3' terminal end of the larger ribosomal RNA. This effect is fairly specific and was observed in preparations of ribosomes and 60S ribosomal subunits from both wheat germ and yeast, although rat liver polysomes did not respond (Schindler and Davies, 1977). The larger RNA of the 50S ribosomal subunit of E. coli is also cleft, although the toxin is a poor inhibitor of translation by E. coli ribosomes. However, cleavage of E. coli ribosomes requires a rather high concentration of alpha sarcin (Schindler and Davies, 1977).

Enomycin is a basic protein (MW = 11,000), produced by Streptomyces mauvecolor, which is a potent inhibitor of polypeptide synthesis in cell-free systems from Ehrlich ascites carcinoma cells, rat liver (Mizuno, Nitta and Umezawa, 1967a), and yeast (Battaner and Vázquez, 1971). Enomycin inhibits binding of aminoacyl-tRNA to polysomes and preliminary experiments suggest that it interacts with both 60S and 40S ribosomal subunits (Mizuno, Nitta and Umezawa, 1967b). In reticulocyte lysates enomycin allows formation of initiation complexes on 80S ribosomes and 40S subunits. In fact these complexes accumulate in the presence of the inhibitor (Mizuno and Umezawa, 1976), since the EF-1-dependent binding of aminoacyl-tRNA is prevented by enomycin (Fernández-Puentes and Vázquez, 1977). Enomycin does not inhibit peptide bond formation in model systems although the initial dipeptide (f-Met-Val) is not formed in reticulocyte lysates in the presence of this inhibitor (Mizuno and Umezawa, 1976). Furthermore, although enomycin also blocks the formation of the EF-2-GTP-80S-ribosome complex it does not affect translocation, suggesting that the toxin cannot interact with pretranslocated polysomes (Fernández-Puentes and Vázquez, 1977).

Phenomycin is a basic protein (MW = 10,000) produced by Streptomyces fervens that specifically prevents protein synthesis in eukaryotic cells (Nishimura, 1968a). In addition it blocks polypeptide synthesis in crude cell-free systems from both rabbit reticulocytes and rat liver (Nishimura, 1968b). In reticulocyte lysates, phenomycin does not prevent formation of initiation complexes on 80S ribosomes and such complexes do in fact accumulate in the presence of the inhibitor. However it has been reported that phenomycin blocks formation of the initial dipeptide (Yamaki et al., 1974). This is compatible with an inhibitory effect of the toxin on EF-1-dependent binding of aminoacyl-tRNA; this possibility remains to be tested.

Modeccin, a protein present in the root of Adenia digitata (Modecca digitata), is very toxic to mammalian cells by blocking protein synthesis (Refsnes et al., 1977; Stirpe et al., 1978). The inhibitory effect of the toxin in reticulocyte cell-free systems is considerably enhanced by 2-mercaptoethanol, but dissociation of the toxin into subunits decreases the toxicity to eukaryotic cells similarly as with abrin and ricin. Furthermore HeLa cells treatment with neuraminidase increases their sensitivity to modeccin, whereas lactose treatment decreases cell sensitivity to the toxin by preventing its binding to cell receptors (Refsnes et al., 1977; Gasperi-Campani et al., 1978). Modeccin inhibits in cell-free systems the formation of the EF-2-GTP-ribosome complex and blocks EF-1-dependent binding of aminoacyl-tRNA but does not affect the non-enzymic binding of aminoacyl-tRNA to ribosomes (Montanaro et al., 1978).

2.1.7 Chartreusin

Chartreusin is an antibiotic produced by *Streptomyces chartreusis* which is active on gram-positive bacteria, mycobacteria, and eukaryotic cells but has little activity on gram-negative bacteria (Leach et al., 1953). The chemical structure of chartreusin has already been established (Eisenhuth, Stamm and Schmid, 1964) (Fig. 19).

Most of the studies on the mode of action of chartreusin have been carried out in eukaryotic systems. Chartreusin might interfere with EF-1 interaction with the ribosome since the enzymic binding is more sensitive to

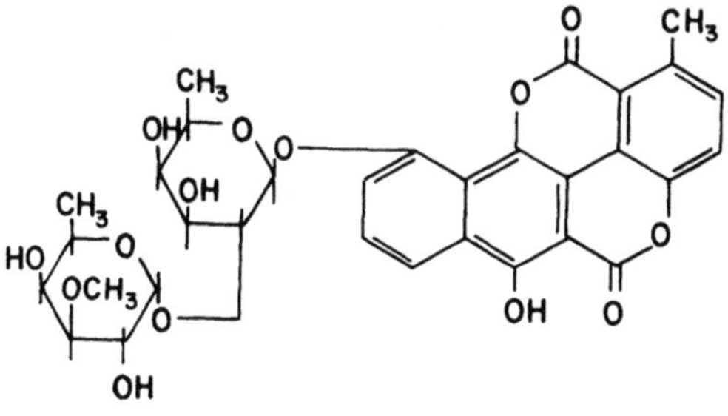

$C_{32}H_{32}O_{14}$; MW 640.58

Fig. 19. Chemical structure of chartreusin

chartreusin than the non-enzymic binding and no effect was detected on EF-1-dependent GTPase in preliminary experiments (Gregg and Heintz, 1972). However, to a certain extent chartreusin also inhibits non-enzymic binding of Phe-tRNA (Gregg and Heintz, 1972) and Ac-Phe-tRNA to mammalian ribosomes (Carrasco, Fernández-Puentes and Vázquez, 1976).

2.1.8 Aminochromes and Catechols

A number of aminochromes (Goodwin, Shafritz and Weissbach, 1969) and catechols (homogentisic acid and related compounds) (Peterson, Raghupathy and McKean, 1971) have been reported to inhibit poly(U)-directed phenylalanine incorporation by mammalian cell-free systems. Adrenochrome (Fig. 7) is also active in E. coli cell-free systems at the ribosome level (Shafritz, Goodwin and Weissbach, 1969). Adrenochrome inhibits EF-2-dependent binding of aminoacyl-tRNA to human tonsil ribosomes (Carrasco and Vázquez, 1972). However it is not yet known whether this inhibition is due to an effect of adrenochrome on the codon-anticodon interaction. Adrenochrome, therefore, besides inhibiting step (d), might also be an inhibitor of step (b) of the initiation phase (see above). Furthermore, adrenochrome inhibits formation of the ternary complex eIF-2-Met-tRNA$_F$-GTP in initiation in eukaryotic systems, as indicated above.

2.1.9 Heparin

Heparin is a mucopolysaccharide with strong inhibitory effects on elongation in bacterial cell-free systems and on both initiation and elongation in eukaryotic systems. For the sake of simplicity we will refer to all of these effects in this chapter. The inhibitory effect of heparin on polyphenylalanine synthesis in bacterial cell-free systems appears to be due to the strong negative charge of the compound, which probably causes a chelation of magnesium ions, since it can be reversed by the addition of polycations (Wacker, Ishimoto and Chandra, 1967; Berlinguet and Normand, 1968). In eukaryotic cell-free systems heparin has been reported to inhibit binding of Phe-tRNA to rat lymphosarcoma ribosomes dependent on EF-1 and to 40S subunits dependent on an unresolved factor (Koka and Nakamoto, 1968). Further work has indeed shown that EF-1 interacts strongly with heparin and although this mucopolysaccharide does not interfere with the formation of the EF-1·GTP·aminoacyl-tRNA complex, it totally blocks its interaction with the ribosome, thus preventing step (d) in translation (Slobin, 1976). An even stronger inhibitory effect of heparin has been observed in initiation in a Krebs ascites cell-free system (Waldman and Goldstein, 1973). This effect is specific to heparin, other mucopolysaccharides tested being inactive; furthermore the structural requirements for heparin to have this effect are distinct from those necessary for its anticoagulant effect (Waldman, Marx and Goldstein, 1974; Goldstein, Waldman and Marx, 1975). The inhibitory effect of heparin is due to the strong interaction with some component(s) of the crude initiation factor preparation used in these studies (Goldstein, Waldman and Marx, 1975; Waldman, Marx and Goldstein, 1975).

2.1.10 TPCK (1-chloro-4-phenyl-3-tosylamido-2-butanone

The compound 1-chloro-4-phenyl-3-tosylamido-3-butanone (TPCK) (MW = 351.85) inhibits protein synthesis in HeLa cells, virus-transformed 3T3 mouse fibroblasts and mouse plasmacytoma cells (Pong, Nuss and Koch, 1975). TPCK does not inhibit amino acid incorporation in a reticulocyte cell-free system for elongation of the polypeptide chain and therefore does not inhibit EF-1-dependent aminoacyl-tRNA binding to ribosomes (Highland et al., 1974). However,TPCK causes polysome breakdown in intact cells, suggesting that it blocks peptide chain initiation (Pong, Nuss and Koch, 1975), although these results have not been confirmed in resolved cell-free systems. On the other hand in bacterial cell-free systems TPCK strongly inhibits polypeptide elongation by blocking EF-T-dependent binding of aminoacyl-tRNA (Sedlacek, Jonak and Rychlik, 1971; Jonak, Sedlacek and Rychlik, 1971, 1973). TPCK does not inhibit formation of the EF-Tu·GTP complex (Sedlacek, Jonak and Rychlik, 1971) but inactivates EF-Tu, preventing formation of the EF-Tu·GTP·aminoacyl-tRNA complex (Richman and Bodley, 1973). Furthermore TPCK has also been reported to block the stringent factor- and ribosome-dependent synthesis of magic spots I and II by an unknown mechanism (Jonak and Clark, 1973).

2.2 Compounds Causing Misreading

All the compounds in this group affect the ribosomal conformation at the level of the small ribosomal subunit. We include here a widely represented group of aminoglycoside antibiotics having in common (a) either a streptamine or deoxystreptamine or epistreptamine moiety in their structure, (b) a cationic structure, (c) a broad spectrum of antibacterial activity, (d) a substantial misreading of the genetic code resulting in the incorporation of amino acids in some positions which do not correspond to the coding nucleotide triplets, and (e) a bactericidal effect which is not yet well explained since most of the other antibacterial inhibitors of translation are bacteriostatic.

According to the sugar moiety we can distinguish within this group of antibiotics those compounds which contain streptamine (streptomycin, dihydrostreptomycin, hybrimycin A_1, hybrimycin A_2 and hybrimycin A_3), epistreptamine (hybrimycin B_1, hybrimycin B_2, hybrimycin B_3), and deoxystreptamine (neamine, neomycin B, neomycin C, kanamycin A, kanamycin B, kanamycin C, tobramycin, gentamicin A, gentamicin C_{1a}, sisomicin, gentamicin C_1, gentamicin C_2, paromomycin II, ribostamycin, butirosin A, butirosin B, lividomycin A, lividomycin B and amikacin (Figs. 20-29).

Of the antibiotics that contain deoxystreptamine, a number have the deoxyparomamine or paromamine (D-glucosamine linked to 2-deoxystreptamine) moiety (lividomycin B, paromomycin, gentamicin A and kanamycin C). They differ from the other misreading aminoglycosides in being active not only in bacterial but also in eukaryotic ribosomes.

Within the group of aminoglycoside antibiotics there are a number that have no effect on misreading (kasugamycin and spectinomycin), or an inhibitory effect on another step might be more relevant for the mode of action (hygromycin B). Therefore these antibiotics will not be considered here.

It is widely accepted that the characteristic deoxystreptamine or epistreptamine or streptamine moiety is a common and absolute requirement for the biological activity of the aminoglycoside antibiotics (Tanaka, Masukawa and Umezawa, 1967; Masukawa and Tanaka, 1968; Davies, 1970). However, these sugars are not antibacterial by themselves and have to be linked to one or more sugars with at least one amino group to permit biological activity of the antibiotic. The attachment of different amino groups or sugars to this moiety in well defined positions can drastically affect the antibiotic activity of the aminoglycosides (Benveniste and Davies, 1973b).

Since all the above aminoglycoside antibiotics are of a cationic nature they have a number of effects in intact bacteria and their cellular components. Thus streptomycin and dihydrostreptomycin can interact with nucleic acids and proteins, being effective reagents for the precipitation of DNA, RNA, ribosomes, and casein. Streptomycin and dihydrostreptomycin can also interfere with cell membrane function in permeability and the oxidation of certain substrates. However, we will not discuss these effects of the antibiotics here since they are not related to their inhibitory effects on protein synthesis.

Besides the aminoglycoside antibiotics, a number of compounds and incubation conditions have been found to cause misreading in bacterial cell-free systems, including polyamines, some organic solvents, increasing concentrations of Mg^{++} ions, and suboptimal temperatures of incubation. It is very interesting that a high fidelity in translation of the eukaryotic ribosomes has been observed, since no significant misreading by mammalian systems was detected in the presence of any of the compounds and under the conditions indicated above. In fact the only compounds which have been described as being effective in causing misreading in mammalian systems are the aminoglycoside antibiotics that contain the deoxyparomamine or the paromamine moiety, and hygromycin B.

2.2.1 Streptomycin and Dihydrostreptomycin

Streptomycin was initially discovered in cultures of *Streptomyces griseus* (Schatz, Bugie and Waksman, 1944) and further obtained from a number of *Streptomyces* species. The chemical structure of streptomycin was soon elucidated (Lemieux and Wolfrom, 1948). Since streptomycin was the first non toxic antibiotic which has a broad antibacterial spectrum and is even active on *Mycobacterium*, it has acquired great importance in medical use over the last thirty years. Dihydrostreptomycin, obtained by hydrogenation of streptomycin, is also used therapeutically and in biological studies as an inhibitor of translation (Fig. 20).

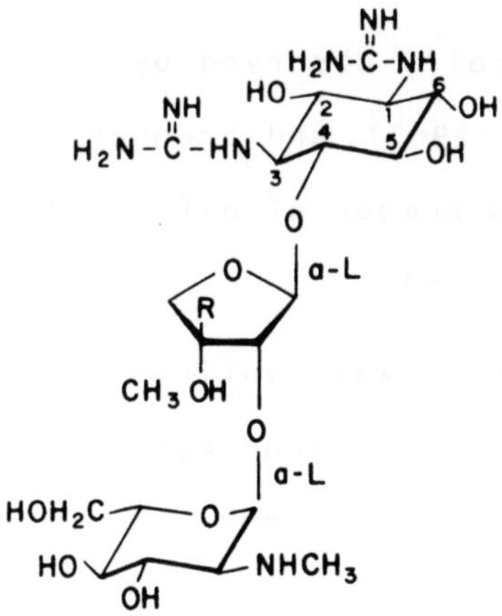

Streptomycin R= -CHO ($C_{21}H_{39}O_{12}N_7$; MW 581.87)

Dihydrostreptomycin R= -CH_2OH ($C_{21}H_{41}O_{12}N_7$; MW 583.88)

Fig. 20. Chemical structures of streptomycin and dihydrostreptomycin

After early studies showed an inhibitory effect on protein synthesis by streptomycin in intact bacteria, it was shown that the antibiotic also prevents polypeptide synthesis by crude bacterial cell-free systems (Erdös and Ullmann, 1959). On the basis of genetic considerations it was postulated that streptomycin-sensitivity, -resistance and -dependence are expressed in the ribosome, which is the site of action of the antibiotic (Spotts and Stanier, 1961). Indeed streptomycin causes the incorporation of amino acids in intact bacteria in some positions which do not correspond to the coding nucleotide triplets (misreading), and the presence of the antibiotic results in the growth of some streptomycin-dependent mutants in the absence of some essential amino acids (Gorini and Kataja, 1964, 1965). Due to the misreading effect some streptomycin-sensitive auxotrophs are also able to grow with sub-lethal concentrations of streptomycin in the absence of some essential amino acids, and this phenomenon is now called phenotypic repair or phenotypic suppression, since the phenotype is altered by the antibiotic (Gorini and Kataja, 1965). Suppression of amber and ochre nonsense mutations by streptomycin was initially observed by Benzer and Champe (1962), Garen and Siddiki (1962), and Brenner and Beckwith (1965) and later described in a number of different bacterial systems (for a review see Davies, 1966).

The misreading effect of streptomycin was confirmed in bacterial cell-free systems resulting in the polynucleotide-directed incorporation of some amino acids which are not incorporated in the absence of the antibiotic (Davies, Gilbert and Gorini, 1964). It has been postulated that the misreading effect by aminoglycoside antibiotics could be related to their association effect on ribosomes (Spirin, 1972). However, it appears that the ability to stabilize the association of the two ribosomal subunits does not necessarily lead to errors in translation, since it has been observed that in the case of two anomeric aminoglycosides stabilizing 70S ribosomes to a similar extent, only one of them is bactericidal and causes misreading (Cousin et al., 1977). The induction of misreading by streptomycin and other aminoglycoside antibiotics has been confirmed in cell-free systems with natural messengers. Misreading was further shown in these systems at very low streptomycin concentrations which allow some ribosomes to initiate (Tai, Wallace and Davis, 1978).

As earlier postulated by Spotts and Stanier (1961) the effect of streptomycin is located on the bacterial ribosome (Davies, Gilbert and Gorini, 1964; Flaks et al., 1962; Speyer, Lengyel and Basilio, 1962; Mager, Benedict and Artman, 1962) at the 30S subunit (Davies, 1964; Cox, White and Flaks, 1964). One of the proteins of the 30S ribosomal subunit (protein S12) is altered in streptomycin-resistant and streptomycin-dependent mutants (Traub and Nomura, 1968; Birge and Kurland, 1969) and leads to resistant and dependent ribosomes respectively in reconstitution experiments. The resistant mutants have different point alterations in the protein S12; thus, mutants have been studied in which (a) lysine has been replaced by either asparagine, threonine, or arginine in position 42 (Funatsu and Wittmann, 1972; Funatsu, Nierhaus and Wittmann, 1972), (b) lysine has been replaced by arginine in position 87 (Funatsu and Wittmann, 1972; Funatsu, Nierhaus and Wittmann, 1972), and (c) one tyrosine has been replaced by phenylalanine and one valine by one alanine in positions not determined (Brakier-Gingras,

Lacoste and Boileau, 1974). Protein 30B is altered in Bacillus subtilis mutants resistant to streptomycin (Itoh et al., 1975); this protein is homologous with protein S12 of E. coli. On the other hand reversion from streptomycin dependence to independence has been observed, leading to new mutants in which either protein S4 or protein S5 is altered (Hasenbank et al., 1973; Birge and Kurland, 1970; Olsson, Isaksson and Kurland, 1974). The alterations in protein S5 are single amino acid replacements (Itoh and Wittmann, 1973; Piepersberg et al., 1975) whereas alterations in protein S4 are more drastic and protein chains shorter or longer than the wild type S4 are formed (Funatsu, Nierhaus and Wittmann, 1972; Hasenbank et al., 1973). Furthermore, there are a number of E. coli streptomycin-resistant and streptomycin-dependent mutants in which the three proteins S4, S5, and S12 are altered and the phenotypes are similar to streptomycin-resistant and streptomycin-dependent mutants in which only protein S12 is altered (Wittmann and Apirion, 1975).

Besides causing misreading, streptomycin also prevents elongation of the polypeptide chain (Modolell and Davis, 1969; Luzzatto, Apirion and Schlessinger, 1968, 1969). This takes place by causing premature fall off of streptomycin-bound ribosomes (Wallace et al., 1973). Therefore the interaction of streptomycin with the ribosome has no immediate inhibitory consequences in initiation but protein synthesis is prevented by a cyclic blockade, since those streptomycin-bound ribosomes are impaired in IF-3-dependent dissociation and become engaged in an abortive initiation and elongation falling off again from the initiation complexes or from polysomes in elongation (Wallace et al., 1973; Wallace, Tai and Davis, 1973). Streptomycin also blocks termination by preventing recognition of the release factors RF-1 and RF-2 (Scolnick et al., 1968).

All the evidence suggest that the inhibitory effects of streptomycin on initiation, elongation, and termination are pleiotropic effects due to the same interaction(s) of the antibiotic with the ribosome. The interaction of streptomycin with the 30S subunit appears to be a rather controversial and complex process. For binding studies streptomycin is frequently replaced by dihydrostreptomycin, because of conveniency for the preparation of the radioactive labeled compound, since both antibiotics appear to have the same target. (^{3}H) dihydrostreptomycin binds to sensitive ribosomes (Wolfgang and Lawrence, 1967) and the 30S subunit from sensitive (Kaji and Tanaka, 1968; Vogel et al., 1970; Ozaki, Mizushima and Nomura, 1969) but not resistant ribosomes (Ozaki, Mizushima and Nomura, 1969). One single binding site (Chang and Flaks, 1972; Schreiner and Nierhaus, 1973) and two binding sites (Kaji and Tanaka, 1968; Biswas and Gorini, 1972) on the ribosome have been proposed for the binding of the streptomycin antibiotics. It has further been reported that the two binding sites of the antibiotic are located on the 16S RNA (Garvin, Biswas and Gorini, 1974). These apparently contradictory reports appear to be due, at least partly, to the use of different monovalent ion concentrations in the binding experiments. Thus at low, but not at high, NH_4^+ concentrations the antibiotic interacts with the 16S RNA and also with the 23S RNA (Schreiner and Nierhaus, 1973). Binding experiments have been performed with 30S subunits reconstituted from protein deficient cores containing only 16S RNA and the proteins S4, S7, S8, S15, S16, S17, S19 (Schreiner and Nierhaus, 1973). These cores do not bind dihydrostreptomycin, although their binding capacity can be restored by addition of proteins S3 + S5 + S9 + S10 + S14, but the binding is independent of the addition of protein S12 (Schreiner and Nierhaus, 1973). On the other hand two peaks of radioactivity were found in affinity labeling of ribosomes with a streptomycin analog: one of the peaks contained proteins S3 and S4 and the other peak contained proteins S14, S18, and S19 (Pongs and Erdmann, 1973). These results are compatible with those obtained in the above report in reconstitution experiments (Schreiner and Nierhaus, 1973) but appear to conflict with the results obtained in binding experiments with a (^{3}H)-labeled photo-activated analog of streptomycin since proteins S7, S14 and S16/S17 were labeled (Girshovich, Bochkareva and Ovchinnikov, 1976).

Some information concerning the binding site of the streptomycin is also available from results obtained studying the inhibition by different proteins of the antibiotic interaction with the ribosomes. Thus dihydrostreptomycin binding to the 30S ribosomal subunit is strongly inhibited by the specific anti-S19 protein immunoglobulin G fragment (Lelong et al., 1974), which adds weight to the importance of this protein in the streptomycin interaction site as shown by other independent methods indicated above (Schreiner and Nierhaus, 1973; Pongs and Erdmann, 1973). Dihydrostreptomycin binding was affected to a lesser extent by antibodies against proteins S1, S10, S11, and S21 but only weakly by the anti-S12 immunoglobulin G fragment (Lelong et al., 1974). Dihydrostreptomycin binding to ribosomes can be prevented by a mixture of initiation factors, although the antibiotic does not appear to compete directly with those factors (Lelong et al., 1972).

The nature of the primary effect of streptomycin on the ribosome is not well understood, but the antibiotic certainly causes a rearrangement or distortion affecting a number of components of the small subunit. Thus streptomycin reactivates inactive ribosomes and inactivates active ribosomes for Phe-tRNA binding (Miskin and Zamir, 1972). Furthermore, ribosomes of a streptomycin-resistant mutant, owing to a replacement of the lysine in position 87 of protein S12, are very active for non-enzymic translocation, whereas parachloromercuri-benzoate treatment of the protein is required to activate the non-enzymic translocation in the wild strain (Asatryan and Spirin, 1975). Moreover streptomycin treatment of the ribosomes considerably enhances the reactivity with kethoxal of RNA and protein of both 30S and 50S ribosomal subunits (Delihas, Topol and Larrinúa, 1975). Conformational changes of ribosomal structure caused by streptomycin treatment have been proposed (Sherman and Simpson, 1969; Brakier-Gingras, Provost and Dugas, 1974; Delihas, Topol and Larrinúa, 1975). Indeed the interaction of streptomycin with the ribosome modifies the iodination of at least five ribosomal proteins; whereas iodination by antibiotic binding is decreased in the case of proteins S10 and S12 + L20, it is increased in two proteins of the larger ribosomal subunit, namely L17 and L28 (Martinez, Vãzquez and Modolell, 1978). Furthermore the antibiotic paromomycin does not interact with 30S ribosomal subunits unless 50S subunits or streptomycin are present. These results suggest that interaction of 50S subunits or streptomycin with the 30S subunits favor a conformational change required for the binding of paromomycin (Lando et al., 1976).

2.2.2 Other Aminoglycoside Inducers of Misreading

The neomycin complex of antibiotics was obtained from culture filtrates of Streptomyces fradiae (Waksman and Lechevalier, 1949), and the neomycins A, B, and C were resolved (Swart, Lechevalier and Waksman, 1951). Neomycin A was identified as neamine (Dutcher and Donin, 1952) and the chemical structures of neomycin B, neomycin C, and neamine were later established (Rinehart et al., 1962; Hichens and Rinehart, 1963) (Fig. 21).

The complex antibiotic kanamycin was initially discovered in culture filtrates from Streptomyces kanamyceticus (Umezawa et al., 1957). Three different compounds A, B, and C were isolated from the antibiotic complex (Umezawa et al., 1957; Cron et al., 1958; Schmitz et al., 1958) and the chemical structures of kanamycin A (Cron et al., 1958), kanamycin B (Ito, Motohiro and Osawa, 1964) and kanamycin C (Murase, 1961) were elucidated (Fig. 22).

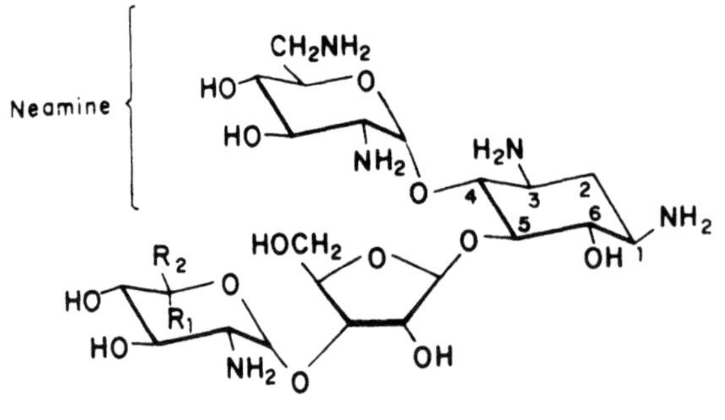

Neamine (synonym neomycin A) ($C_{12}H_{26}O_6N_4$; MW 322.36)

Neomycin B : R_1 = $-CH_2-NH_2$; R_2 = $-H$ ($C_{23}H_{46}O_{13}N_6$; MW 614.64)

Neomycin C : R_1 = $-H$; R_2 = $-CH_2-NH_2$ ($C_{23}H_{46}O_{13}N_6$; MW 614.64)

Fig. 21. Chemical structures of the neomycins

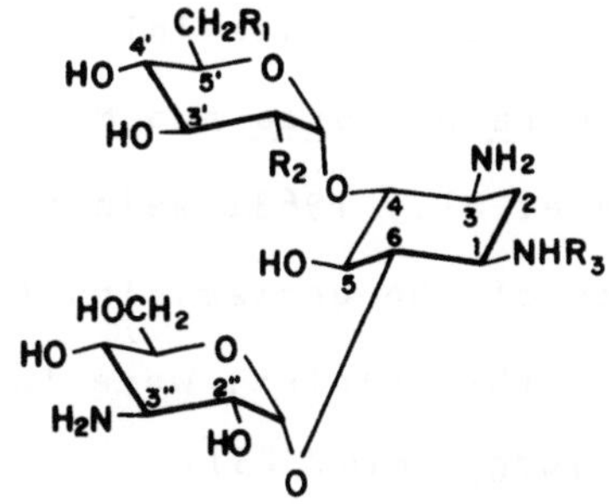

Kanamycin A : R_1 = -NH_2 ; R_2 = -OH ; R_3 = -H ($C_{18}H_{36}O_{11}N_4$; MW 484.50)

Amikacin : R_1 = -NH_2 ; R_2 = -OH ; R_3 = -CO-CHOH-CH_2-CH_2-NH_2 ($C_{22}H_{43}O_{12}N_5$; MW 585.60)

Kanamycin B : R_1 = -NH_2 ; R_2 = -NH_2 ; R_3 = -H ($C_{18}H_{37}O_{10}N_5$; MW 483.51)

Kanamycin C : R_1 = -OH ; R_2 = -NH_2 ; R_3 = -H ($C_{18}H_{36}O_{11}N_4$; MW 484.50)

Fig. 22. Chemical structures of amikacin and the kanamycins

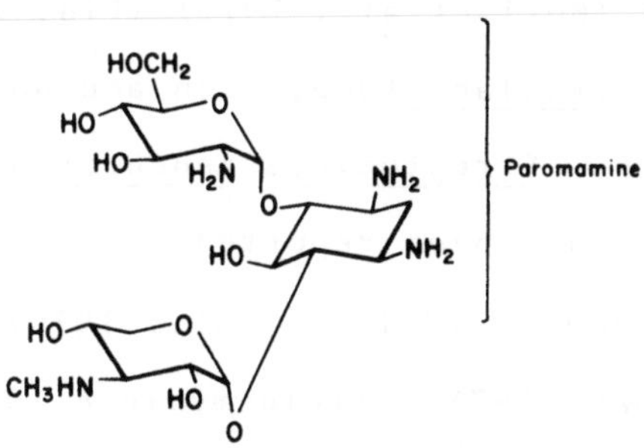

Gentamicin A ($C_{18}H_{36}O_{10}N_4$; MW 468.50)

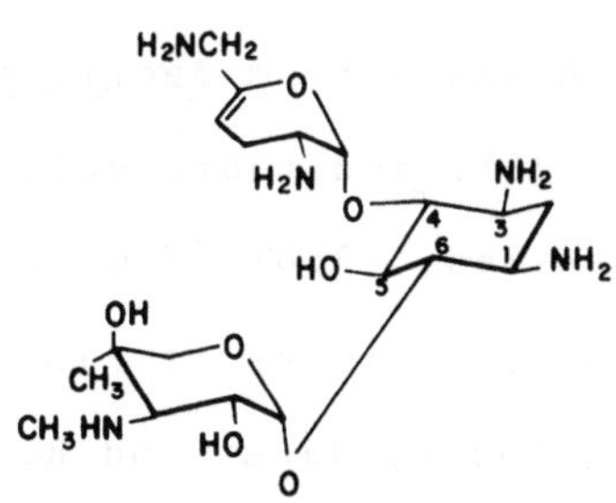

Sisomicin ($C_{19}H_{37}O_7N_5$; MW 447.55)

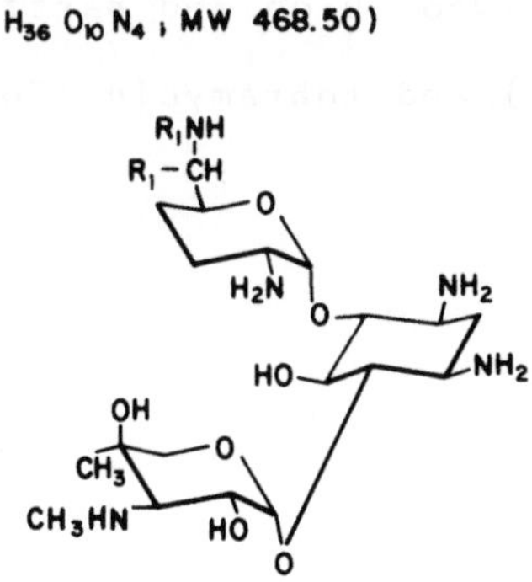

Gentamicin C_1 : R_1 = -CH_3 ; R_2 = -CH_3 ($C_{21}H_{43}O_7N_5$; MW 477.59)

Gentamicin C_2 : R_1 = -CH_3 ; R_2 = -H ($C_{20}H_{41}O_7N_5$; MW 463.56)

Gentamicin C_{1a} : R_1 = -H ; R_2 = -H ($C_{19}H_{39}O_7N_5$; MW 449.54)

Fig. 23. Chemical structures of sisomicin and gentamicins

The gentamicin complex was initially obtained from cultures of Micromonospora purpurea and Micromonospora echinospora (Weinstein et al., 1963; Weinstein et al., 1964). The chemical structures of the gentamicins A, C_{1a}, C_1 and C_2, the major components of the complex, were later established (Maehr and Schaffner, 1970) (Fig. 23).

Many other aminoglycoside antibiotics have been developed in the last ten years. A number of the them are obtained from cultures of different microorganisms. Thus the obtention of ribostamycin from Streptomyces ribosidificus (Shomura et al., 1970) (Fig. 24), sisomicin from Micromonospora myoensis (Fig. 23) (Cooper, Jaret and Reiman, 1971), lividomycins A and B from Streptomyces lividus (Mori et al., 1971) (Fig. 25), butirosins A and B from Bacillus circulans (Woo, Dion and Bartz, 1971) (Fig. 26), and tobramycin from Streptomyces tenebrarius (Wick and Welles, 1968) (Fig. 27) have been reported. The chemical structures of ribostamycin (Akita et al., 1970), sisomicin (Cooper, Jaret and Reiman, 1971), lividomycin A (Oda et al., 1971), butirosins A and B (Woo, Dion and Bartz, 1971), paromomycins (Daniels et al., 1971), and tobramycin (Koch and

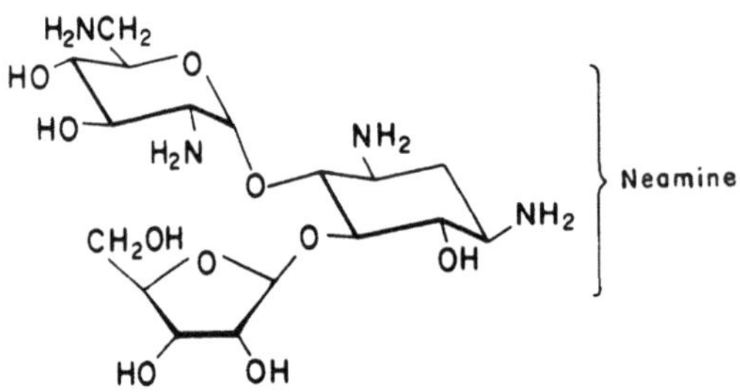

$C_{17}H_{34}O_{10}N_4$; MW 454.49

Fig. 24. Chemical structure of ribostamycin

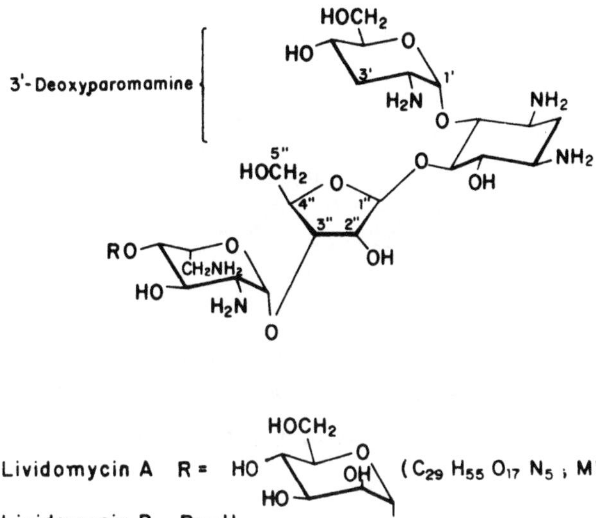

Lividomycin A R = ($C_{29}H_{55}O_{17}N_5$; MW 745.77)

Lividomycin B R = -H

Fig. 25. Chemical structure of the lividomycins

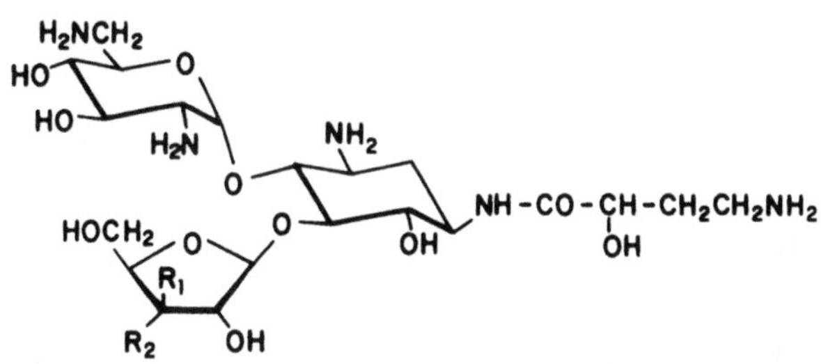

Butirosin A : R_1 = -OH ; R_2 = -H ($C_{21}H_{41}O_{12}N_5$; MW 555.57)

Butirosin B : R_1 = -H ; R_2 = -OH ($C_{21}H_{41}O_{12}N_5$; MW 555.57)

Fig. 26. Chemical structures of butirosins A and B

Rhoades, 1968) were soon elucidated (Figs. 23-28). Hybrimycins A1, A2, and A3 are obtained when a Streptomyces fradiae mutant is grown in the presence of streptamine, whereas hybrimycins B1, B2, and B3 are obtained when the mutant is grown in the presence of epistreptamine (Shier, Rinehart and Gottlieb, 1969, 1970) (Fig. 29). On the other hand a number of semisynthetic

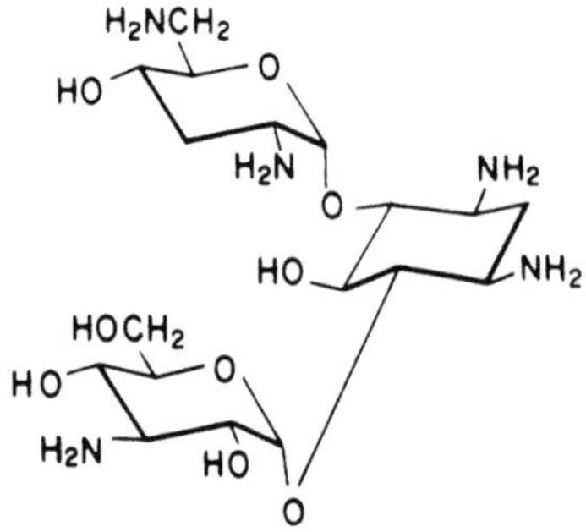

$C_{18}H_{37}O_9N_5$; MW 467.51

Fig. 27

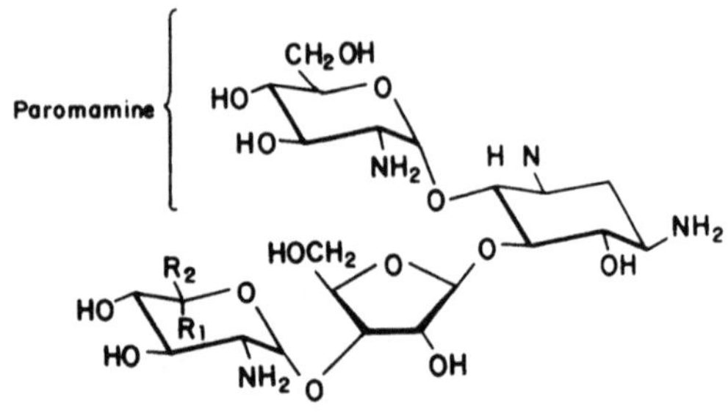

Paromomycin I : R_1 =-CH_2NH_2 ; R_2 =-H

Paromomycin II : R_1 =-H ; R_2 =-CH_2NH_2

$C_{23}H_{45}O_{14}N_5$; MW 615.65

Fig. 28

Fig. 27. Chemical structure of tobramycin

Fig. 28. Chemical structures of the paromomycins

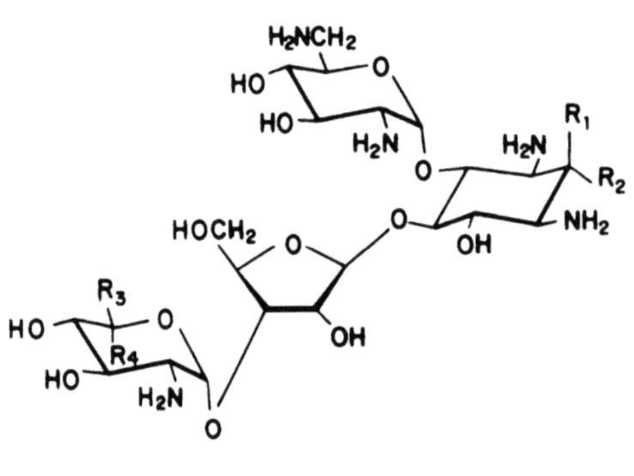

$C_{23}H_{46}O_{14}N_6$, MW 630.64

Hybrimycin A1 : R_1 =-H ; R_2 =-OH ; R_3 =-H ; R_4 =-CH_2NH_2

Hybrimycin A2 : R_1 =-H ; R_2 =-OH ; R_3 =-CH_2NH_2 ; R_4 =-H

Hybrimycin B1 : R_1 =-OH ; R_2 =-H ; R_3 =-H ; R_4 =-CH_2NH_2

Hybrimycin B2 : R_1 =-OH ; R_2 =-H ; R_3 = CH_2NH_3; R_4 =-H

Fig. 29. Chemical structures of the hybrimycins

aminoglycoside antibiotics have been prepared, the best known of which is amikacin (Kavaguchi et al., 1972) (Fig. 22).

All the antibiotics described above which have been tested induce misreading in intact bacteria, since phenotypic suppression has been observed in E. coli auxotrophs grown in the absence of some essential amino acids but in the presence of these antibiotics, as observed with streptomycin (Gorini and Kaṭaja, 1964 , 1965). In bacterial cell-free systems they block R17 phage RNA-directed polypeptide synthesis (Benveniste and Davies, 1973b) and cause misreading in different systems (Benveniste and Davies, 1973b; Davies, Gorini and Davis, 1965; Davies and Davis, 1968; Lando, Cousin and Garilhe, 1973). However these antibiotics do not appear to have the same target site on the ribosome as streptomycin. Thus it has been proposed that streptomycin exerts a single-site action on the ribosome but that the other aminoglycosides have a multi-site action (Davies and Davis, 1968). This would explain the difficulty in finding aminoglycoside-resistant mutants, other than streptomycin, in which the protein altered has been identified; furthermore, when this was achieved it was not the same protein as in the case of streptomycin-resistant mutants (Tanaka, Sashikata and Umezawa, 1967). Thus streptomycin resistance is due to a mutation of protein S12 (Traub and Nomura, 1968; Birge and Kurland, 1969), whereas there are two separate loci for neamine resistance (Cannon, Cabezôn and Bollen, 1974). The mutational alteration for one of these loci is on protein S17 (Bollen et al., 1975) and involves the replacement of histidine by proline in position 30 of protein S17 (Yaguchi and Wittmann, 1976).

Surprisingly protein L6 of the larger ribosomal subunits is altered in a gentamicin-resistant mutant (Buckel et al., 1977).

Differences in the site of action of the streptomycin antibiotics and other aminoglycoside antibiotics are also supported by binding studies with the labeled antibiotics. Thus binding of dihydrostreptomycin is inhibited by streptomycin, enhanced by neomycin, paromomycin, and lividomycin and not affected by kanendomycin (kanamycin B sulphate), butirosins, gentamycin, and kanamycin (Lando et al., 1976). On the other hand paromomycin binding to ribosomes is enhanced by dihydrostreptomycin and streptomycin, inhibited by neomycin, lividomycin, kanendomycin, butirosins and not affected by kanamycin (Lando et al., 1976). Interestingly paromomycin does not bind to 30S ribosomal subunits unless 50S subunits or streptomycin are present to induce on the 30S the appropriate change in conformation (Lando et al., 1976).

Aminoglycoside antibiotics containing either the deoxyparomamine or paromamine moiety (lividomycin B, paromomycin, gentamicin A, and kanamycin C) are also active in intact cells of _Tetrahymena thermophila_. They also inhibit translation and cause misreading by eukaryotic ribosomes even at low concentrations. Neamine, gentamicin A, gentamicin C_{1a}, kanamycin A, and kanamycin 1A also cause misreading by eukaryotic ribosomes, but only at rather high concentrations (Palmer and Wilhelm, 1978; Wilhelm, Pettitt and Jessop, 1978).

2.2.3 Miscellaneous Compounds and Conditions Causing Misreading

A significant increase in misreading by _E. coli_ cell-free systems has been observed in the presence of Mg^{++} at supraoptimal concentrations (Davies, Gilbert and Gorini, 1964; Szer and Ochoa, 1964), of a number of aliphatic polyamines such as spermine and spermidine (Friedman and Weinstein, 1964), and of some organic solvents, namely ethanol (So and Davie, 1964; So, Bodley and Davie, 1964). Furthermore, it was observed in cell-free systems from the thermophilic bacteria Bacillus stearothermophilus that there is a minimum of misreading at 65°C, which corresponds to the optimal growth temperature, and the misreading is increased when either the temperature is lowered or the Mg^{++} concentration is increased (Weinstein, Friedman and Ochoa, 1966). On the other hand neither of the compounds and conditions indicated here can cause significant misreading in cell-free systems from mammalian ribosomes (Weinstein, Friedman and Ochoa, 1966).

3. Inhibitors of Peptide Bond Formation [Step (e)]

3.1 Puromycin and Analogs

The antibiotic puromycin was isolated from culture filtrates of Streptomyces alboniger (Porter et al., 1952) and the chemical structure was established by degradative studies and total synthesis (Fig. 30) (Waller et al., 1953; Baker et al., 1955). Puromycin is a very wide spectrum antibiotic and it is active on gram-positive and gram-negative bacteria, amebae, and protozoa, but has hardly been used clinically since it is also very active on mammalian cells. The antibiotic homocitrullylamino adenosine, which is closely related to puromycin (Fig. 30), has been obtained from mycelia of Cordyceps militaris (Kredich and Guarino, 1961).

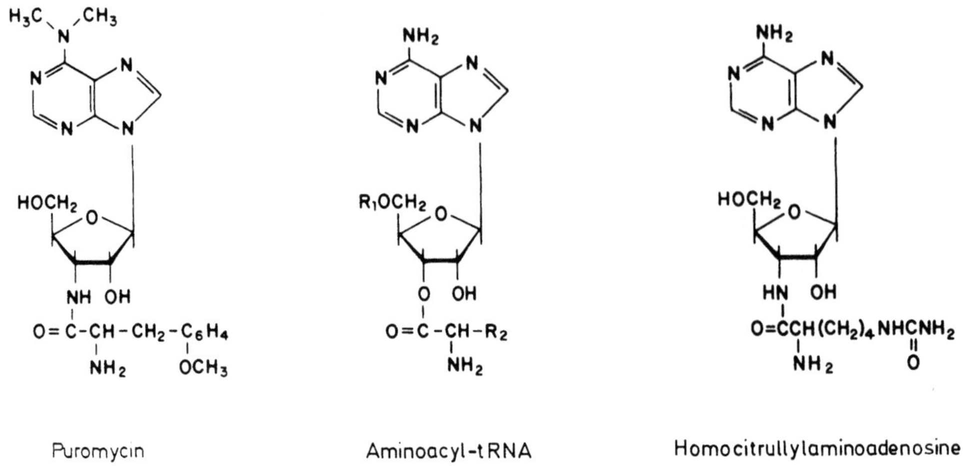

$C_{22}H_{29}O_5N_7$; MW 471.52

Fig. 30. Chemical structures of puromycin and homocitrullyl-aminoadenosine

The great importance of puromycin lies in its being the clearest example of an antibiotic inhibitor of translation whose chemical structure is obviously an analog of a component of protein synthesis. Indeed puromycin is an analog of the 3' terminal end of aminoacyl-tRNA (Yarmolinsky and Haba, 1959) (Fig. 30) and therefore interacts with the acceptor site of the ribosomal peptidyl transferase center on the larger ribosome subunit (50S or 60S). However, puromycin lacks that part of the aminoacyl-tRNA molecule responsible for interactions with the ribosome other than with the peptidyl transferase. Hence, the use of the antibiotic provides a simplified method for the study of peptide bond formation in a reaction in which the $-NH_2$ group of puromycin becomes linked to the C-terminal end of f-Met or the peptidyl group ("puromycin reaction"). The product of the puromycin reaction (f-Met- or peptidyl-puromycin) is unable to take part in the next step of protein synthesis. However, all the evidence indicates that the formation of a peptide bond between puromycin and the f-Met- or peptidyl-group takes place by the same mechanism as peptide bond formation in protein synthesis (Traut and Monro, 1964; Monro, Maden and Traut, 1967). Polyphenylalanyl-tRNA (Traut and Monro, 1964; Monro, Maden and Traut, 1967), polylysyl-tRNA (Rychlik, 1966; Gottesman, 1967), f-Met-$tRNA_F$ (Zamir, Leder and Elson, 1966; Bretscher and Marcker, 1966), and Ac-Phe-tRNA (Haenni and Lucas-Lenard, 1968) are suitable donor substrates in the puromycin reaction. The terminal fragments CACCA-, AACCA-, ACCA- and CCA-Met-f from f-Met-$tRNA_F$ (and also CACCA-Phe-Ac and CACCA-Leu-Ac) undergo a ribosome-catalyzed reaction with puromycin to yield f-Met-puromycin in a simplified system known as the fragment reaction. The reaction requires only the larger ribosomal subunit, monovalent and divalent cations, and either methanol or ethanol, and also takes place with the corresponding Ac-Phe-oligonucleotides from the 3' terminal end of Ac-Phe-tRNA (Monro et al., 1969; Vázquez et al., 1969). The limiting factor in this reaction is the affinity of the substrate for the donor site of the peptidyl transferase; thus peptide bond formation also takes place with the donor substrates CA- and A-Met-f, but a high concentration of A-Met-f is required since its affinity for the active center is very low (Černá et al., 1973a, 1973b). Thus the donor site of the peptidyl transferase appears to have specific sites not only for adenosine but also for cytidylic acid, since this compound increases the binding of A-Met-f considerably (over 400%) (Černá et al., 1976). The structural requirement for activity as an acceptor substrate like puromycin has been studied by using a number of puromycin derivatives and aminoacyl-nucleosides (Nathans and Neidle, 1963;

Waller et al., 1966; Harbon and Chapeville, 1970; Harris, Hanlon and Symonds, 1971; Rychlik et al., 1969; Rychlik et al., 1970; Pestka et al., 1973; Pozdnyakov et al., 1972; Duquette, Ritter and Vince, 1974; Gooch and Hawtrey, 1975; Ariatti and Hawtrey, 1975; Žemlička et al., 1975). In cell-free systems Phe-, Tyr-, and Leu-adenosine equal puromycin in activity as acceptor substrates in the peptide bond formation reaction (Waller et al., 1966; Harbon and Chapeville, 1970; Harris, Hanlon and Symonds, 1971; Rychlik et al., 1969, 1970). The affinity of puromycin for polysomes (Pestka, 1972a) is considerably higher than for ribosomes (Fernández-Muñoz and Vázquez, 1973b). The puromycin derivatives α-hydroxy- and demethoxy-α-hydroxypuromycin are of special significance. These compounds lack the α-amino group of puromycin and are therefore unable to form a peptide bond, but they form an ester bond with the f-Met moiety of f-Met-tRNA or CACCA-Met-f in a reaction catalyzed by the peptidyl transferase center of the ribosome (Fahnestock et al., 1970). The kinetics of the puromycin reaction have been studied in the "fragment reaction" assay and in a polysomal system (Fernández-Muñoz and Vázquez, 1973a; Pestka, 1972a). The weak interaction of puromycin with the acceptor site of the peptidyl transferase center is inhibited by Phe-adenosine (Fernández-Muñoz and Vázquez, 1973b).

In order to identify the ribosomal components involved in the interaction with puromycin, affinity labeling studies have been carried out using a number of puromycin analogs and derivatives. Unfortunately the results obtained in these studies appear to be rather contradictory and do not clarify the interaction site of puromycin. Thus (^{3}H)N-bromoacetyl puromycin specifically interacts with the 50S subunit but the (^{3}H)N-bromoacetyl puromycin nucleoside that is inactive as an acceptor substrate in peptide bond formation interacts to a greater extent than the active compound (Minks, Ariatti and Hawtrey, 1975). N-iodoacetyl puromycin has been reported to interact with protein L6 of bacterial ribosomes but the compound interacts to a great extent with protein S18 although it is known that puromycin acts on the 50S ribosomal subunit (Pongs et al., 1973). Treatment of rat liver ribosomes with N-iodoacetyl-puromycin leads to labeling of two proteins of the 60S subunits (Stahl, Dressler and Bielka, 1974) (proteins L27 and L28 in Bielka's nomenclature, which are most probably proteins L28 and L29 in the nomenclature adopted here). On the other hand the labeling of the puromycin analog 5'-O-(N-bromoacetyl-p-aminophenyl-phosphoryl)3'-N-L-phenylalanyl-puromycin aminonucleoside bound to bacterial ribosomes was attained by reaction with bacterial ribosomes with N-acetyl-L-(^{3}H)leucyl-pentanucleotide bound, since the N-acetyl-L-(^{3}H)-leucine moiety is transferred to the puromycin analog; with this method there is no labeling of ribosomal proteins, only the 23S RNA being labeled, which suggests its involvement in the acceptor site of the peptidyl transferase center (Greenwell, Harris and Symonds, 1974). In these affinity labeling experiments the pentanucleotide sequence G-U-C*-C-Gp of 23S RNA has been isolated, where C* represents C modified by the affinity label (Eckermann and Symonds, 1978). (^{3}H)puromycin labels protein L23 of bacterial ribosomes by irradiation at 253.7 nm; there is also some labeling of protein S14 but this appears to be non-specific since it is also observed as a result of irradiation by N-(ethyl-

2-diazomalonyl)-(^3H)puromycin, which is rather inactive (Cooperman et al., 1975). We do not really know the meaning of these results but other studies suggest that protein L16 of bacterial ribosomes (Ballesta et al., 1974; Moore et al., 1975; Bernabeu, Vázquez and Ballesta, 1977) and proteins L21, L26, L31 and L36 of rat liver ribosomes (Reyes, Vázquez and Ballesta, 1976a, 1976b, 1977; Czernilofsky et al., 1977) may be involved in the peptidyl transferase center.

There are few reports concerning the mode of action of homocitrullylaminoadenosine, but all the evidence so far obtained suggests that this antibiotic acts in a similar way to puromycin in cell-free extracts of both *Escherichia coli* and rat liver (Guarino, Ibershof and Swain, 1963).

3.2 Chloramphenicol and Analogs

Chloramphenicol is produced in the fermentation broth of *Streptomyces venezuelae* (Ehrlich et al., 1947). The chemical structure was early established (Fig. 31) and confirmed by synthesis (Controulis, Rebstock and Crooks, 1949) and X-ray diffraction studies (Dunitz, 1952). Since chloramphenicol has a wide antibacterial spectrum, being active on gram-positives and gram-negatives, and has a remarkable permeability in cells and tissues it has been widely used clinically. In view of this and the simplicity of its molecule chloramphenicol was the first synthetic antibiotic commercialized.

There are two asymmetric carbon atoms in the chloramphenicol molecule and four stereoisomers were therefore synthesized, of which only the D(-)threo stereoisomer, that is

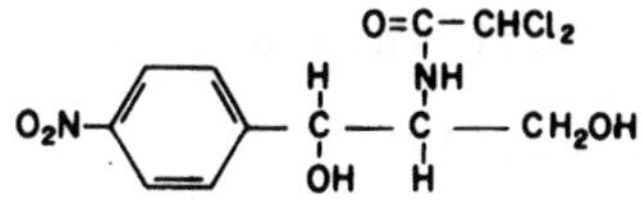

$C_{11}H_{12}O_5N_2Cl_2$; MW 323.14

Chloramphenicol

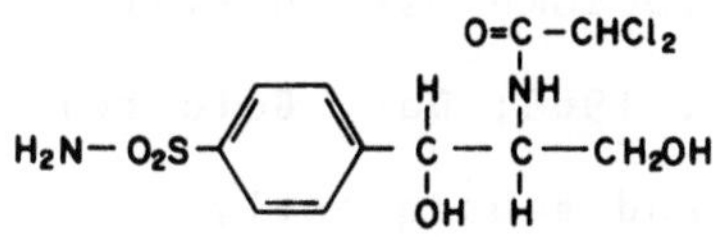

$C_{11}H_{12}O_4N_2S$, MW 268.28

D-AMP-3

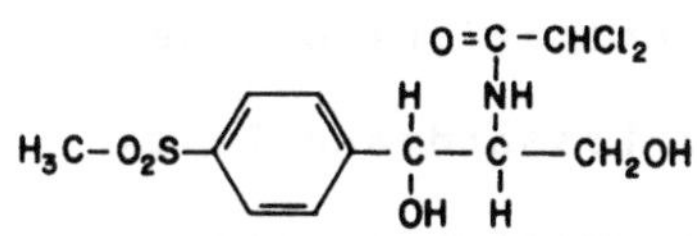

$C_{12}H_{15}O_5NSCl_2$; MW 356.23

Thiamphenicol

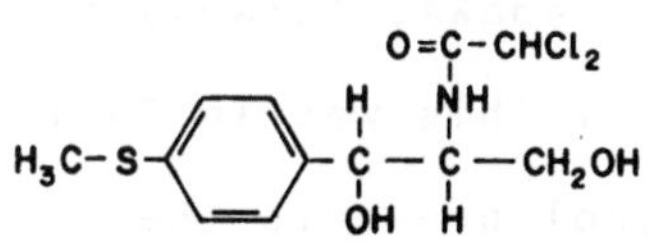

$C_{12}H_{15}O_3NCl$, MW 256.70

D-Win 5094

Fig. 31. Chemical structures of chloramphenicol and analogs

precisely the natural antibiotic, has significant antibacterial activity. However, there are a number of chloramphenicol analogs with antibacterial action, such as thiamphenicol, Win 5094 and AMP-3 (Fig. 31) but in all cases the D(-)threo configuration is required.

Chloramphenicol inhibits protein synthesis in prokaryotic type cells and organelles, and the extent of inhibition in cell-free systems depends on the synthetic polynucleotide used as an artificial mRNA (Speyer et al., 1963; Kučan and

Lipmann, 1964; Vázquez, 1966b). Chloramphenicol interacts with ribosomes of bacteria (Vázquez, 1963; Vázquez, 1966a), chloroplasts,and blue-green algae (Anderson and Smillie, 1966; Rodríguez-López and Vázquez, 1968) on the 50S ribosomal subunit (Vázquez, 1964a). One binding site on the ribosome was initially proposed for the antibiotic (Wolfe and Hahn, 1965; Das, Goldstein and Kanner, 1966; Hurwitz and Braun, 1968) and subsequently demonstrated at the inhibitory concentrations (Fernández-Muñoz et al., 1971; Fernández-Muñoz and Vázquez, 1973a), but more binding sites can be observed at higher concentrations of the antibiotic (Das, Goldstein and Kanner, 1966; Lessard and Pestka, 1972). Chloramphenicol blocks peptide bond formation in intact bacteria (Cundliffe and McQuillen, 1967) and bacterial cell-free systems (Goldberg and Mitsugi, 1967a; Coutsogeorgopoulos, 1967; Vogel, Vogel and Elson, 1971; Pestka, 1970a; Tanaka, Teraoka and Tamaki, 1971; Černá, Jonák and Rychlik, 1971). This was in fact observed in many systems, since chloramphenicol prevents the reaction of either polyphenylalanyl-tRNA (Traut and Monro, 1964; Maden and Monro, 1968), polylysyl-tRNA (Rychlik, 1966; Gottesman, 1967; Goldberg and Mitsugi, 1967a; Coutsogeorgopoulos, 1967; Pestka, 1970a; Tanaka, Teraoka and Tamaki, 1971; Černá, Jonák and Rychlik, 1971),or dipeptidyl-tRNA (Černá, Jonák and Rychlik, 1971) with puromycin. However, the most conclusive evidence for the inhibitory effect of chloramphenicol on the peptidyl transferase center has been obtained in the fragment reaction (Monro, 1967; Monro and Vázquez, 1967), since it is the most fully resolved assay for peptide bond formation. The inhibitory effect of chloramphenicol in this system has been reported to be mixed competitive for puromycin (Fernández-Muñoz and Vázquez,

1973a), which is in agreement with previous reports on the puromycin reaction (Goldberg and Mitsugi, 1967a), although there are other hypotheses (Coutsogeorgopoulos, 1967; Pestka, 1972b). Chloramphenicol has no effect on substrate binding (CACCA-Leu-Ac) to the donor site of the peptidyl transferase center (Monro and Vázquez, 1967) but it inhibits binding of both puromycin (Fernández-Muñoz and Vázquez, 1973b) and aminoacyl-oligonucleotides to the acceptor site of the active center (Monro et al., 1970; Celma, Monro and Vázquez, 1970, 1971; Černá and Rychlik, 1972; Pestka, 1969a, 1969b, 1969c). 2'-O-aminoacyl-dinucleoside phosphates are good inhibitors of chloramphenicol binding and their interaction site appears to overlap with that of the antibiotic, though this is not so with the corresponding 3'-O-aminoacyl-dinucleoside phosphates (Goldberg, Ringer and Chládek, 1977). From reconstitution experiments on the 50S subunit from protein deficient cores it has been concluded that ribosomal protein L16 is involved in chloramphenicol binding (Nierhaus and Nierhaus, 1973). A similar conclusion was drawn from affinity labeling of ribosomes in vitro with monoiodo-(^{14}C)-amphenicol since protein L16 is labeled (Pongs, Bald and Erdmann, 1973) although proteins S6, L16, and L24 are predominantly labeled when affinity labeling experiments are performed with intact bacteria (Pongs and Messer, 1976). Furthermore the loss of protein L16 from the ribosomes closely parallels the loss of activity of the particles to bind (^{14}C)chloramphenicol, implying an important role for this protein in the interaction of the antibiotic with the ribosome (Bernabeu, Vázquez and Ballesta, 1977). On the other hand in affinity labeling of bacterial ribosomes with (^{14}C)bromamphenicol, proteins L2 and L27 are preferentially labeled (Sonenberg, Wilckek and Zamir, 1973). (^{14}C)chloramphenicol interacts with bacterial ribosomes and polysomes (Pestka, 1974). In this case however there is a different affinity of the antibiotic for the ribosomes depending on the site of interaction of the substrates. Thus chloramphenicol has a similar affinity for free ribosomes and ribosomes with peptidyl-tRNA bound to the A-site, but the affinity for the antibiotic is increased when the substrate is translocated to the P-site (Contreras and Vázquez, 1977a). This is precisely the situation in which chloramphenicol interacts with the polysomes to block translation. Chloramphenicol and some active analogs (thiamphenicol, p-SO_2-NH_2, and p-S-CH_3 analogs) are mutually exclusive in their binding to the ribosome although in some aspects they differ in their mode of action (Contreras, Barbacid and Vázquez, 1974). (^{14}C)thiamphenicol, like chloramphenicol, increases its affinity for the ribosome when peptidyl-tRNA is bound to the donor site (Contreras and Vázquez, 1977a).

Although chloramphenicol has a clear and specific action in bacteria as an inhibitor of translation, it has been shown to inhibit DNA synthesis in eukaryotic cells at very high concentrations (Freeman, Patel and Haldar, 1977) and this effect appears to be mainly responsible for the irreversible toxicity of the antibiotic in certain cases in medical use (various contributions in Postgraduate Medical Journal, 1974).

3.3 Antibiotics of the Lincomycin Group

Lincomycin was initially obtained from cultures of *Streptomyces lincolnensis* (Mason, Dietz and Deboer, 1963) and the chemical structure was soon elucidated (Hoeksema et al., 1964) (Fig. 32).

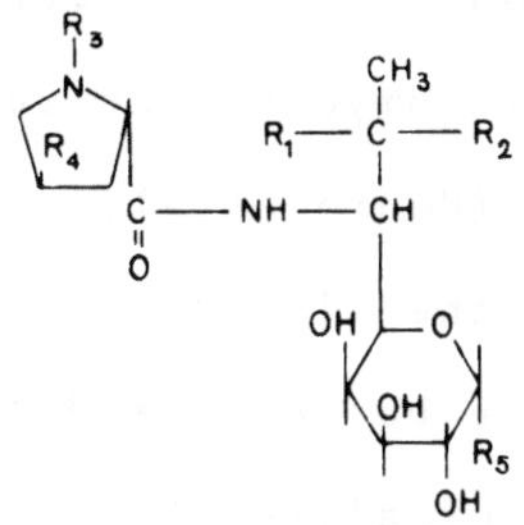

	Compound	Substituent at position				
		R_1	R_2	R_3	R_4	R_5
A	Lincomycin	OH	H	CH_3	C_3H_7	SCH_3
B	4'-Depropyl-4'-ethyllincomycin (U 21.699)	OH	H	CH_3	C_2H_5	SCH_3
C	Lincomycin sulfoxide	OH	H	CH_3	C_3H_7	S-CH_3 O
D	1-Demethylthio-1-hydroxylincomycin	OH	H	H	C_3H_7	OH
E	7-Chloro-7-deoxy-N-demethyl-4'-depropyl-4'-pentyllincomycin	H	Cl	H	C_5H_{11}	SCH_3
F	7-Chloro-7-deoxy-N-demethyl-4'-depropyl-4'-hexyllincomycin	H	Cl	H	C_6H_{13}	SCH_3
G	7-Chloro-7-deoxy-N-demethyl-lincomycin	H	Cl	H	C_3H_7	SCH_3
H	7-Chloro-7-deoxy-N-demethyl-N-ethyl-4'-depropyl-4'-butyllincomycin	H	Cl	C_2H_5	C_4H_9	SCH_3
J	7-Chloro-7-deoxylincomycin (clinimycin or clindamycin)	H	Cl	CH_3	C_3H_7	SCH_3
K	7-Chloro-7-deoxylincomycin (epi-clinimycin or epi-clindamycin)	Cl	H	CH_3	C_3H_7	SCH_3
L	7-Chloro-7-deoxy-4'-depropyl-4'-ethyllincomycin	H	Cl	CH_3	C_2H_5	SCH_3
M	4'-Depropyl-4'-penthyl-N-demethyllincomycin	OH	H	H	C_5H_{11}	SCH_3
N	N-Demethyllincomycin	OH	H	H	C_3H_7	SCH_3
P	7-O-Acetatelincomycin	OH	C_2H_3O	CH_3	C_3H_7	SCH_3
Q	Celesticetin	OCH_3	H	CH_3	H	SC_2H_4-O-C 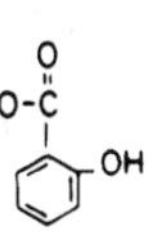

Lincomycin ($C_{18}H_{34}O_6N_2S$; MW 406.53)

Celesticetin ($C_{24}H_{36}O_9N_2S$; MW 528.61)

Clindamycin ($C_{18}H_{33}O_5N_2SCl$; MW 424.98)

Fig. 32. Antibiotics of the lincomycin group

Closely related to lincomycin is the antibiotic celesticetin produced by Streptomyces caelestis (De Boer et al., 1955) in mixture with some related antibiotics including celesticetin B, C,and D, whose structures have also been established (Hoeksema, 1964; Argoudelis and Brodasky, 1972). A number of lincomycin derivatives are also obtained from the cultures of the producer organisms; furthermore many compounds with antibiotic properties were obtained by chemical modification of lincomycin, including the clinically useful clindamycin (Fig. 32) (Monro et al., 1971; Lewis, 1974; reviews). Antibiotics of the lincomycin group are mainly active against gram-positive bacteria and to a lesser extent against gram-negative bacteria,although clindamycin has antiplasmodial activity and is moderately active in gram-negative bacteria as well.

Antibiotics of the lincomycin group (lincomycin, celesticetin,and clindamycin) inhibit protein synthesis in bacteria and bacterial cell-free systems (Vázquez, 1966b; Josten and Allen, 1964; Chang, Sih and Weisblum, 1966), their interactions being with the 50S ribosomal subunit (Vázquez, 1966a; Fernández-Muñoz et al., 1971; Chang, Sih and Weisblum, 1966). Lincomycin affinity for gram-negative bacterial ribosomes is very low, but this affinity is considerably increased in the presence of ethanol (Fernández-Muñoz et al., 1971). One molecule of the antibiotic interacts per ribosome interfering with peptide bond formation (Monro and Vázquez, 1967), preventing binding of the 3' terminal end of substrates to the donor (Celma, Monro and Vázquez, 1970; Černá and Rychlik, 1972; Pestka, 1969a) and to the acceptor (Pestka, 1969a; Celma, Monro and Vázquez, 1971) sites of the peptidyl transferase center. Indeed lincomycin inhibits binding to the ribosome of peptide bond formation inhibitors, such as chloramphenicol (Vâzquez, 1966a). Conversely different peptide bond formation inhibitors block the binding of lincomycin to ribosomes (Fernândez-Mũnoz et al., 1971). Furthermore protein L16 appears to be required in these reactions, since the activity of ribosomes for lincomycin and chloramphenicol binding and peptidyl transferase activity suffers a parallel decrease when protein L16 is separated by washing the ribosomes with increasing concentrations of NH_4Cl in the presence of ethanol (Bernabeu, Vázquez and Ballesta, 1977).

Lincomycin was also postulated to inhibit initiation (Cundliffe, 1969; Pestka, 1972b; Reusser, 1975), aminoacyl-tRNA binding (Chang, Sih and Weisblum, 1966), and translocation (Burns and Cundliffe, 1973). However the effect of lincomycin on aminoacyl-tRNA binding has only been observed on non-enzymic binding and at high concentrations of the antibiotic (Chang, Sih and Weisblum, 1966). Reports on the effect of lincomycin on initiation are mainly based on the observation that the antibiotic has no effect on polypeptide elongation by polysomes in vitro (Pestka, 1972b; Reusser, 1975) and in vivo (Burns and Cundliffe, 1973) and therefore induces polysomal run-off (Cundliffe, 1969). Indeed, lincomycin effectively inhibit (^{14}C)-chloramphenicol binding to ribosomes (Fernández-Muñoz et al., 1971) but does not affect (^{14}C)chloramphenicol binding to polysomes (Pestka, 1974). Indeed lincomycin does not bind to ribosomes with peptidyl-tRNA bound to either the A- or the P-site (Contreras and Vázquez, 1977a). In fact the antibiotic binds to 50S subunits and free ribosomes and inhibits the early steps of peptide bond formation prior to formation of polysomes

Table 8. Streptogramin antibiotics

Antibiotic complex (synonyms)	Group A (synonyms)	Group B (synonyms)
Ostreogrycin (E 129)	Ostreogrycin A Ostreogrycin C Ostreogrycin D Ostreogrycin G Ostreogrycin Q	Ostreogrycin B Ostreogrycin B_1 Ostreogrycin B_2 Ostreogrycin B_3
Streptogramin	Streptogramin A	Streptogramin B
Virginiamycin (Staphylomycin) (Virgimycin)	Virginiamycin M or M_1 (Staphylomycin M_1) Virginiamycin M_2 (Staphylomycin M_{II})	Virginiamycin S or S_1 (Staphylomycin S) Virginiamycin S_2 Virginiamycin S_3 Virginiamycin S_4
Synergistin (PA-114)	Synergistin A-I (PA-114 A-I)	Synergistin B-I (PA-114 B-I) Synergistin B-3 (PA-114 B-3)
Mikamycin	Mikamycin A	Mikamycin B
Pristinamycin	Pristinamycin II_A	Pristinamycin I_A Pristinamycin I_B Pristinamycin I_C
Vernamycin	Vernamycin A	Vernamycin B_α Vernamycin B_β Vernamycin B_γ Vernamycin B_δ Vernamycin C (Doricin)

Table 8 (continued)

Griseoviridin	Viridogrisein Etamycin Patricin A Patricin B

The antibiotics ostreogrycin A, streptogramin A, virginiamycin M, synergistin A-I, mikamycin A, pristinamycin II_A, and vernamycin A have the same chemical structure. Viridogrisein and etamycin are identical compounds. (Data taken from Vâzquez, 1975b; Crooy and Neys, 1972).

(Contreras and Vázquez, 1977a); this explains why lincomycin allows polysome run-off without affecting initiation. It has also been observed that lincomycin and clindamycin have some effects on initiation complexes formed under certain conditions; the significance of these effects, which were also observed with the 30S ribosomal subunits (Reusser, 1975), is not known, since the main interaction of the antibiotic takes place with the 50S subunit (Fernández-Muñoz et al., 1971; Chang and Weisblum, 1967).

3.4 Antibiotics of the Streptogramin A Group and Griseoviridin

Antibiotics of the streptogramin A group (synonyms mikamycin A, virginiamycin M_1, vernamycin A, ostreogrycin A, pristinamycin II_A, and synergistin A) (Fig. 33) (Table 8) have been reviewed recently (Vázquez, 1975b; Tanaka, 1975). They are antibacterial antibiotics produced by streptomyces, which are mainly active on gram-positive cocci. In most cases streptogramin A antibiotics are produced in mixture with streptogramin B antibiotics, which inhibit translation by a different mechanism (see below).

Streptogramin A antibiotics inhibit protein synthesis in bacteria and bacterial cell-free systems (Vázquez, 1966b, 1966c;Yamaguchi, 1961; Ennis, 1965a, 1965b; Yamaguchi and Tanaka, 1964; Laskin and Chan, 1965) by preventing peptide bond formation, as shown initially in the fragment reaction assay (Monro and Vázquez, 1967) and confirmed in other systems (Pestka, 1970a; Ennis, 1970). These antibiotics interact with the 50S subunit (Vázquez, 1963, 1966a,1967a; Ennis, 1971) and prevent binding of CACCA-Leu-Ac to the donor site (Celma, Monro and Vázquez, 1970; Černá and Rychlik, 1972) and of aminoacyl-oligonucleotides to the acceptor site (Pestka, 1969b; Celma, Monro and Vázquez, 1971) of the peptidyl transferase center. One molecule of these antibiotics is bound per 50S ribosome subunit, protein L16 being an absolute requirement for this binding (Bethune and Nierhaus, 1978). They have also been reported to inhibit EF-T-dependent aminoacyl-tRNA binding under certain conditions (Modolell et al., 1971; Cocito and Kaji, 1971) without affecting GTP hydrolysis (Modolell et al., 1971) and non-enzymic binding of f-Met-$tRNA_F$ (Ennis and Duffy, 1972). There are conflicting reports concerning the effects of these antibiotics on initiation. Streptogramin A antibiotics were proposed as inhibitors of initiation since they have no effect on polypeptide elongation by polysomes (Pestka, 1972b) and induce polysome run-off (Cundliffe, 1969; Ennis, 1972). A certain inhibitory effect on initiation by streptogramin A antibiotics has been reported in a crude system (Tanaka, 1975) but it was not confirmed in a more purified and resolved system studying step by step (Cocito, Voorma and Bosch, 1974). It is known that the lack of effect of streptogramin A on polysomes is due to the low affinity of antibiotic interaction with polysomes (Contreras and Vázquez, 1977a; Ennis, 1974), since peptidyl-tRNA bound to either the A- or the P-site prevents the (^3H)streptogramin A interaction with polysomal ribosomes (Contreras and Vázquez, 1977a). Moreover, streptogramin A antibiotics do not inhibit (^{14}C)chloramphenicol binding to polysomes (Pestka, 1974), although they are very effective in blocking (^{14}C)chloramphenicol binding to free ribosomes (Vázquez, 1963, 1966a). In common with lincomycin, streptogramin A antibiotics interact with 50S ribosomal subunits and free ribosomes prior to polysome formation and inhibit the first few steps of peptide bond formation (Contreras and Vázquez, 1977a).

The antibiotic griseoviridin (Fig. 33) appears to have a mode of action identical to that of streptogramin A (Vázquez,

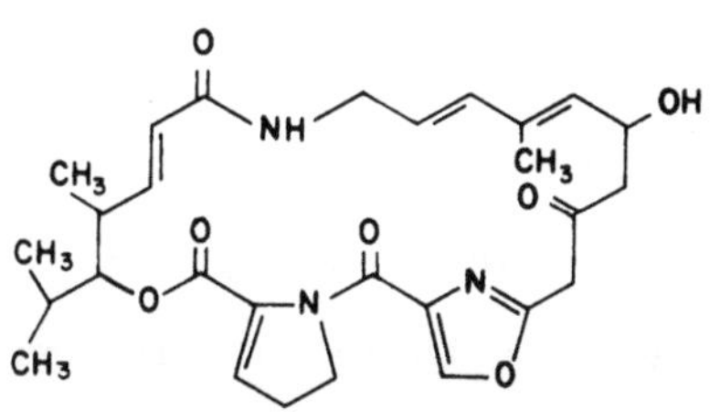

$C_{28}H_{35}O_7N_2$; MW 525.58

Streptogramin A

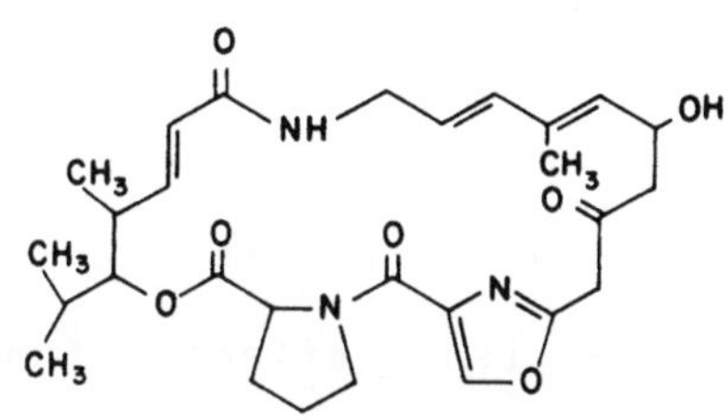

$C_{28}H_{37}O_7N_3$; MW 527.60

Ostreogrycin G

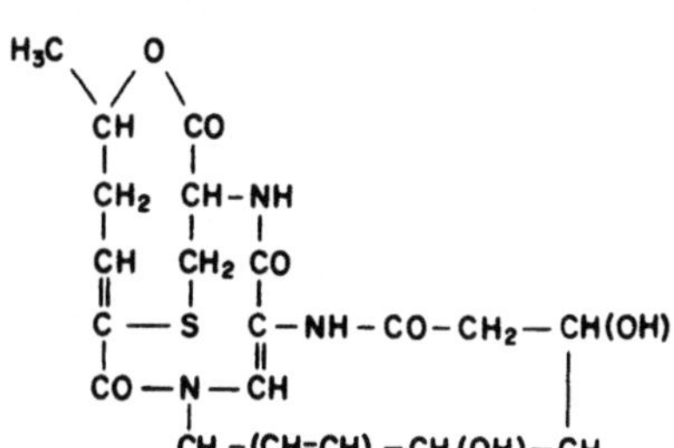

$C_{22}H_{27}O_7N_3S$; MW 477.52

Griseoviridin

Fig. 33. Griseoviridin and antibiotics of the streptogramin A group

1975b). Thus griseoviridin prevents peptide bond formation in bacterial systems (Pestka and Brot, 1971) but has no effect on polysomes (Pestka, 1972b). Although griseoviridin and the streptogramin A antibiotics are antibacterial compounds they appear to have a similar mode of action on the eukaryotic cell-free systems, but their affinity for the eukaryotic ribosome is considerably lower (Barbacid, Contreras and Vázquez, 1975) requiring concentrations one hundred times higher than those required for a similar extent of inhibition in bacterial systems. Although griseoviridin and streptogramin A antibiotics interfere in some way with the interaction of the 3' terminal end of substrates with the donor and acceptor sites of the peptidyl transferase center, their binding appears to be mainly on the P-site since they enhance (^{3}H)gougerotin binding to the A-site (Barbacid, Contreras and Vázquez, 1975; Barbacid and Vázquez, 1974c).

Streptogramin A antibiotics enhance the interaction of streptogramin B antibiotics with the ribosome and this appears to explain the synergistic effects of members of these two groups of compounds (Contreras and Vázquez, 1977b). Plasmid-mediated resistance to streptogramin A antibiotics has been recently shown to be expressed by acetylation of the antibiotic (Goffic et al., 1977).

3.5 Macrolide Antibiotics of the Spiramycin and Carbomycin Groups

The term macrolide has been applied to members of a family of structurally related antibiotics produced by species of streptomyces (Table 9). All macrolide antibiotics contain a large

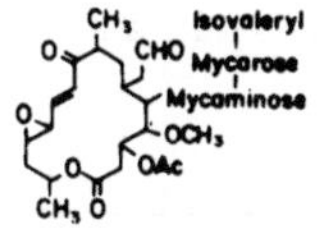

Carbomycin A

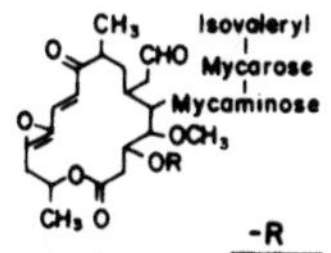

	-R
Carbomycin B :	Acetyl
Niddamycin :	H

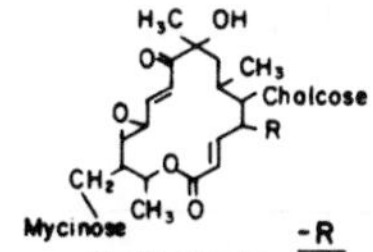

	-R
Chalcomycin :	CH_3
Neutramycin :	H

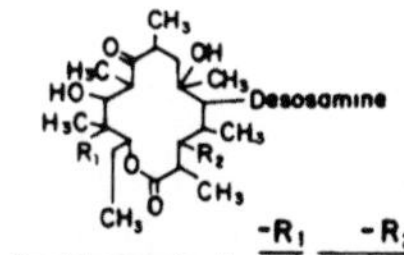

	$-R_1$	$-R_2$
Erythromycin A :	OH	Cladinose
Erythromycin B :	H	Cladinose
Erythromycin C :	OH	Mycarose

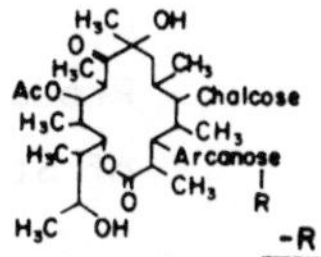

	-R
Lancamycin :	Acetyl
Kujimycin A :	H

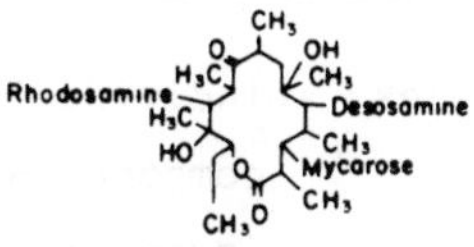

Megalomycin A

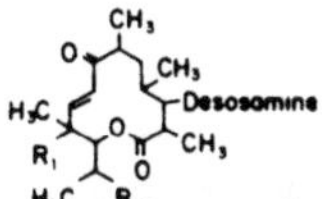

	$-R_1$	$-R_2$
Methymycin :	OH	H
Neomethymycin :	H	OH

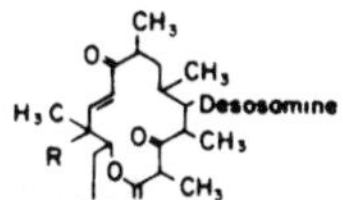

	-R
Narbomycin :	H
Picromycin :	OH

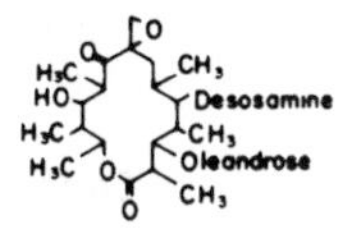

Oleandomycin

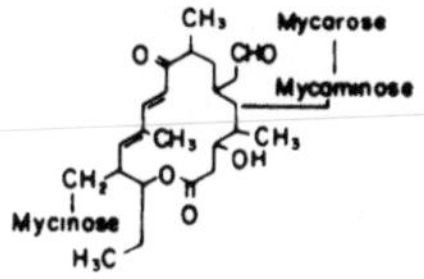

Tylosin

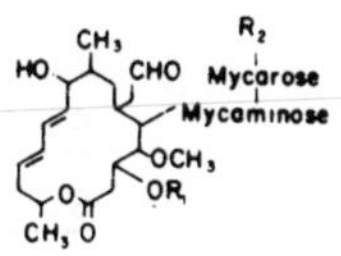

Leucomycins

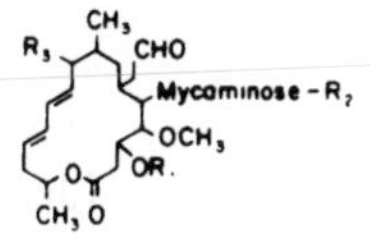

Spiramycins

Carbomycin A	: $C_{42}H_{67}O_{16}N$;	MW 841.96
Carbomycin B	: $C_{42}H_{67}O_{15}N$;	MW 825.96
Niddamycin	: $C_{40}H_{65}O_{14}N$;	MW 783.93
Chalcomycin	: $C_{35}H_{56}O_{14}$;	MW 700.80
Erythromycin A	: $C_{37}H_{67}O_{13}N$;	MW 733.92
Erythromycin B	: $C_{37}H_{67}O_{12}N$;	MW 717.92
Erythromycin C	: $C_{36}H_{61}O_{10}N_2$;	MW 681.87
Lancamycin	: $C_{42}H_{72}O_{16}$;	MW 833.00
Methymycin	: $C_{25}H_{43}O_7N$;	MW 469.60
Narbomycin	: $C_{28}H_{47}O_7N$,	MW 509.66
Picromycin	: $C_{28}H_{47}O_8N$;	MW 525.70
Oleandomycin	: $C_{35}H_{61}O_{12}N$;	MW 687.85
Tylosin A	: $C_{46}H_{77}O_{17}N$;	MW 916.14
Spiramycin I	: $C_{43}H_{74}O_{14}N_2$;	MW 843.04
Spiramycin II	: $C_{45}H_{76}O_{15}N_2$;	MW 885.07
Spiramycin III	: $C_{46}H_{78}O_{15}N_2$;	MW 899.10
Neospiramycin I	: $C_{36}H_{61}O_{10}N_2$;	MW 681.86
Neospiramycin II	: $C_{38}H_{63}O_{11}N_2$;	MW 723.89
Neospiramycin III	: $C_{39}H_{66}O_{11}N_2$;	MW 737.92

	$-R_1$	$-R_2$
A_1 :	H	Isovaleryl
A_3 :	Acetyl	Isovaleryl
A_4 :	Acetyl	n-Butyryl
A_5 :	H	n-Butyryl
A_6 :	Acetyl	Propionyl
A_7 :	H	Propionyl
A_8 :	Acetyl	Acetyl
A_9 :	H	Acetyl

Leucomycin A_3
(synonym Josamycin)

	$-R_1$	$-R_2$	$-R_3$
Spiramycin I :	H	Mycarose	Forosamine
Spiramycin II :	Acetyl	Mycarose	Forosamine
Spiramycin III :	Propionyl	Mycarose	Forosamine
Neospiramycin I :	H	H	Forosamine
Neospiramycin II :	Acetyl	H	Forosamine
Neospiramycin III :	Propionyl	H	Forosamine
Forocidin I :	H	H	H
Forocidin II :	Acetyl	H	H
Forocidin III :	Propionyl	H	H

Fig. 34. Macrolide antibiotics

Table 9. Macrolide antibiotics of the carbomycin and spiramycin groups

(a) Source of antibiotic complexes and the individual components

Antibiotic complex (synonyms)	Individual components (synonyms)	Producer species of Streptomyces
	Angolamycin (Shincomycin A)	*S. eurythermus*
Carbomycin (Magnamycin) (M-4209)	Carbomycin A Carbomycin B	*S. halstedii*, *S. hygroscopicus*, *S. albireticuli*
Leucomycin (Kitasamycin)	Leucomycin A_3 (Josamycin) Leucomycin $A_1, A_4, A_5, A_6, A_7, A_8, A_9$	*S. kitasatoensis* *S. violaceoniger*
	Niddamycin	*S. djakartensis*
	Relomycin	*S. hygroscopicus*
Spiramycin (Rovamycin) (Sequamycin) (Selectomycin) (Provamycin) (Foromacidin)	Spiramycin I Spiramycin II Spiramycin III	*S. ambofaciens*
Tylosin	Tylosin A Tylosin B Tylosin C	*S. hygroscopicus* *S. fradiae*

Data taken from Vázquez (1975a).

Table 9 (continued)

(b) Classification of the sugar moieties linked to the aglycone ring

Group	Characteristics	Members of the group	Sugars linked to the aglycone	
Spiramycin	a	Spiramycins	-mycaminose-mycarose	-isomycamine (synonym forosamine)
		Angolamycin (synonym Shincomycin A)	-angolosamine-mycarose	-mycinose
		Relomycin	-mycaminose-mycarose	-mycinose
		Tylosin	-mycaminose-mycarose	-mycinose
Carbomycin	b	Carbomycins	-mycaminose-mycarose-isovaleryl	
		Josamycin (synonym Leucomycin A_3)	-mycaminose-mycarose-isovaleryl	
		Leucomycins (synonyms Kitasamycins)	-mycaminose-mycarose-R	
		Niddamycin	-mycaminose-mycarose-isovaleryl	
		Turimycin H_0	-mycaminose-mycarose	
		Turimycin H_2	-mycaminose-mycarose-acetyl	

[a]Containing the -mycaminose-mycarose moiety and also another -sugar linked to the aglycone.
[b]Containing the -mycaminose-mycarose-R moiety linked to the aglycone.
Data taken from Vázquez (1975a).

lactone ring with a few or no double bonds and no nitrogen atoms; they have one or more amino sugars, or one non-nitrogenous sugar or both linked to this lactone ring (Figs. 34 and 35).

The mode of action of the macrolide antibiotics has been reviewed recently (Vázquez, 1975a; Oleinick, 1975). They

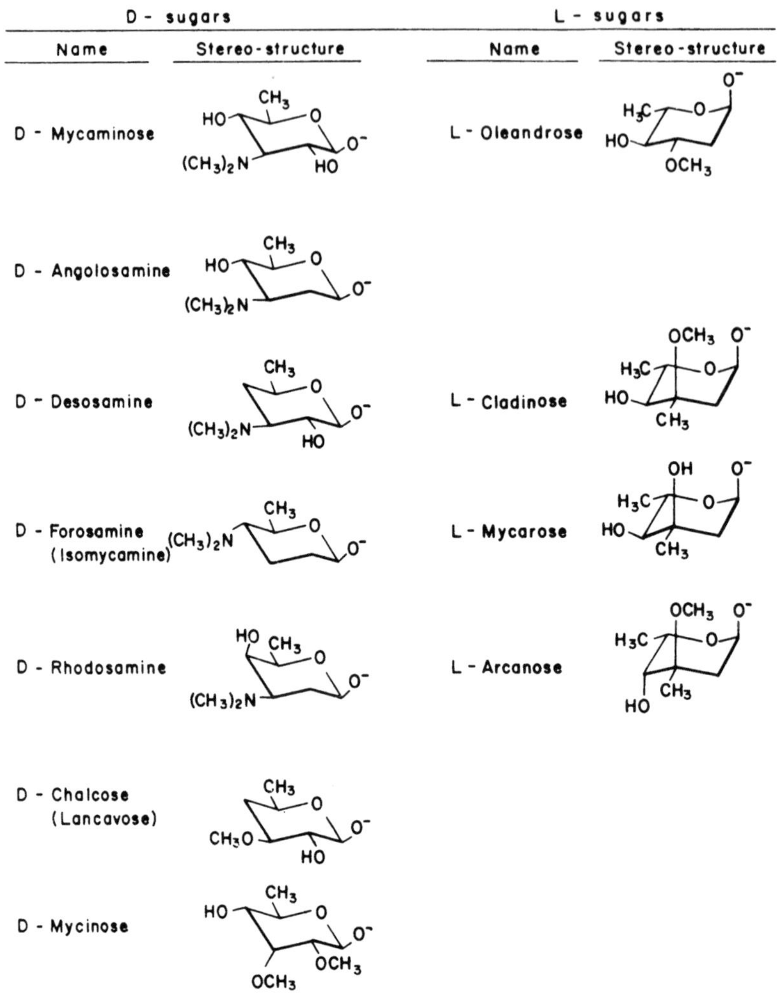

Fig. 35. Sugar moieties from macrolide antibiotics

can be divided according to the sugar moiety linked to the aglycone ring into the spiramycin, carbomycin, erythromycin, methymycin, and lancamycin groups (Vázquez, 1975a). Although all the macrolide antibiotics appear to have a common or overlapping binding site of the aglycone ring to the ribosome, they differ in their mode of action depending on their sugar moiety (Vázquez, 1975a). Thus antibiotics of the spiramycin and carbomycin groups (spiramycins, angolamycin, relomycin, tylosin, carbomycins, josamycins, leucomycins and niddamycin) (Vázquez, 1975a; Haupt et al., 1976) will be considered here. All these antibiotics have in common a -mycaminose-mycarose residue attached to the aglycone ring. The antibiotic spiramycin III blocks protein synthesis in intact bacteria (Vázquez, 1966b; Vázquez, 1967b). There is ample evidence for the interaction of spiramycin III with the larger ribosomal subunit of sensitive bacteria (Vázquez, 1966b, 1967a, 1967b; Ahmed, 1968a, 1968b). Ribosomes from spiramycin-resistant mutants do not bind the antibiotic (Shimizu, Saito and Mitsuhashi; 1970) and have an altered protein L4 (Tanaka, K., et al., 1971). Spiramycin III blocks peptide bond formation in different experimental systems (Monro and Vázquez, 1967; Pestka, 1970a; Černá and Rychlik, 1972; Černá, Rychlik and Pulkrabek, 1969; Tanaka, Teraoka and Tamaki, 1971); it is a rather poor inhibitor of the puromycin reaction when f-Met-$tRNA_F$ is used as the donor substrate but it is a more effective inhibitor when dipeptidyl-tRNA is the donor substrate (Mao and Robishaw, 1971; Kubota, Okuyama and Tanaka, 1972). Binding of both the donor (Celma, Monro and Vázquez, 1970; Černá and Rychlik, 1972) and the acceptor (Pestka, 1969b, Celma, Monro and Vázquez, 1971) substrate to the peptidyl transferase center is inhibited by the antibiotic. However spiramycin III andcarbomycin do not inhibit polypeptide synthesis by polysomes (Pestka, 1972a) and cause polysome breakdown (Cundliffe, 1969; Ennis, 1972). Since macrolide antibiotics do not bind to polysomes (Pestka, 1974) with peptidyl-tRNA bound to either the A- or the P-site of the ribosome (Contreras and Vázquez, 1977a), it appears that spiramycin III acts in a similar way to lincomycin and streptogramin A, inhibiting only the early steps of peptide bond formation prior to formation of polysomes (Contreras and Vázquez, 1977a). All the evidence suggests that the different antibiotics of the spiramycin and carbomycin groups have a mode of action similar to that of spiramycin III (Vázquez, 1963,1966a; Monro and Vázquez, 1967; Černá, Jonák and Rychlik, 1971; Pestka, 1972b; Ennis, 1972; Černá, Rychlik and Pulkrabek, 1969; Mao and Robishaw, 1971, 1972). The -mycaminose-mycarose residue attached to the aglycone ring of these antibiotics appears to be fundamental for their mode of action. Thus neospiramycin III, which is identical to spiramycin III but deprived of the -mycarose residue, is not an inhibitor of peptide bond formation and acts similarly to macrolides of the erythromycin, methymycin and lancamycin groups, none of which has a disaccharide linked to the aglycone ring (Vázquez, 1975a). A similar finding has been observed with demycarosylturimycin (Haupt et al., 1978) whereas turimycin acts like spiramycin III (Haupt et al., 1976).However, all the macrolide antibiotics are mutually exclusive in their binding to the ribosome (Vázquez, 1967a; Wilhelm, Oleinick and Corcoran, 1968; Mao and Putterman, 1969; Fernández-Munoz et al., 1971; Pestka, Nakagawa and Omura, 1974), suggesting a common or overlapping binding site of their aglycone moiety to the 50S ribosomal subunit (Vázquez, 1975a).

3.6 Althiomycin

Althiomycin is an antibacterial antibiotic which is active on gram-positive and gram-negative bacteria, and has been isolated from cultures of Streptomyces althioticus (Yamaguchi et al.,

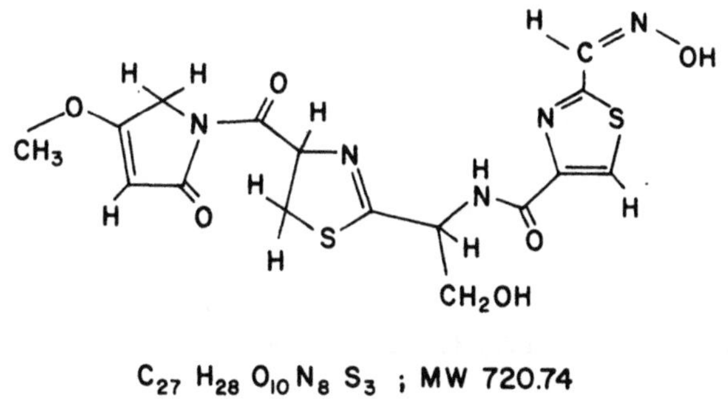

Fig. 36. Chemical structure of althiomycin

1957). The chemical structure of althiomycin has now been elucidated (Fig. 36) (Sakakibara et al., 1974).

Althiomycin inhibits protein synthesis by blocking the peptide bond formation step as shown in the puromycin (Fujimoto et al., 1970) and the fragment reaction assays (B. Cabrer and M.J. San-Millán, unpublished observations), and in the reaction of ribosome-attached peptides with CACCA-Phe (Pestka and Brot, 1971). Althiomycin also blocks peptide bond formation in intact bacteria (Burns and Cundliffe, 1973). Althiomycin is a very effective inhibitor of chloramphenicol binding to polysomes but not to ribosomes (Pestka, 1974).

3.7 Amicetin, Gougerotin, Blasticidin S, Actinobolin, Anthelmycin, and Sparsomycin

Amicetin (synonyms sacromycin and allomycin) is obtained from cultures of *Streptomyces fasciculatus* (McCormick and Hoehn, 1953). The antibiotic is also produced by *Streptomyces plicatus*, in mixture with the closely related antibiotics bamicetin and plicacetin (synonym amicetin B), whose structures have already been elucidated (Haskell et al., 1958; Stevens et al., 1972) (Fig. 37).

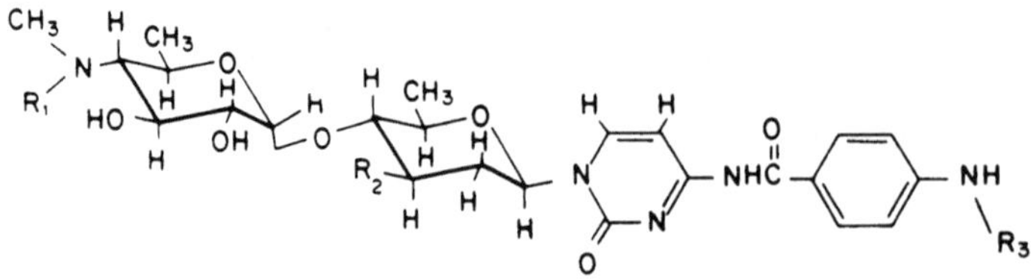

Amicetin: $R_1 = -CH_3$; $R_2 = -H$; $R_3 = -CO-C(CH_3)(NH_2)-CH_2OH$ ($C_{29}H_{42}O_9N_6$; MW 618.71)

Bamicetin: $R_1 = -H$; $R_2 = -H$; $R_3 = -CO-C(CH_3)(NH_2)-CH_2OH$ ($C_{28}H_{40}O_9N_6$; MW 604.64)

Oxamicetin: $R_1 = -CH_3$; $R_2 = -OH$; $R_3 = -CO-C(CH_3)(NH_2)-CH_2OH$ ($C_{29}H_{42}O_{10}N_6$; MW 606.65)

Plicacetin: $R_1 = -CH_3$; $R_2 = -H$; $R_3 = -H$ ($C_{25}H_{35}O_7N_5$; MW 517.56)

Fig. 37. Amicetin group of antibiotics

Closely related to amicetin is also oxamicetin, an antibiotic produced by Arthrobacter oxamicetus (Fig. 37) (Konishi et al., 1973). All these antibiotics are active on gram-positive bacteria and mycobacteria; they have only slight activity on gram-negative bacteria but are active as antitumor agents and in a number of eukaryotic organisms and therefore hold a certain toxicity to humans and animals.

Gougerotin is produced by Streptomyces gougeroti (Kanzaki et al., 1962); its chemical structure has already been established (Kyoichi et al., 1972) (Fig. 38). It is a very broad-spectrum antibiotic active on gram-positive and gram-negative bacteria and a number of eukaryotic organisms; owing to the permeability barrier the antibiotic is less active on mycobacteria, yeast, and fungi.

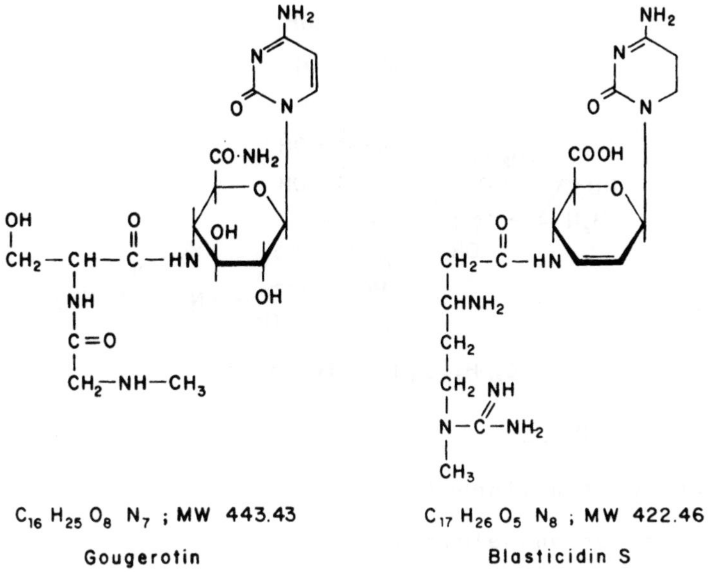

Fig. 38. Chemical structures of gougerotin and blasticidin S

Blasticidin S has been obtained from *Streptomyces globifer* and *S. griseochromogenes* (Takeuchi et al., 1958). Its chemical structure (Fox and Watanabe, 1966) (Fig. 38) and absolute configuration have already been established (Yonehara and Otake, 1966). Blasticidin S is a broad-spectrum antibiotic active on gram-positive and gram-negative bacteria, a number of eukaryotic organisms and as an antitumor agent. The antibiotic is mainly used as an antifungal agent against *Piricularia oryzae* infection of rice fields.

Actinobolin has been isolated from cultures of *Streptomyces griseoviridus* var. *atrofaciens* (Pitillo et al., 1959) and its chemical structure has been elucidated (Munk et al., 1968) (Fig. 39). It is a broad-spectrum antibiotic active on bacteria, eukaryotic organisms, and tumors. It has some limited antimicrobial use in medicine.

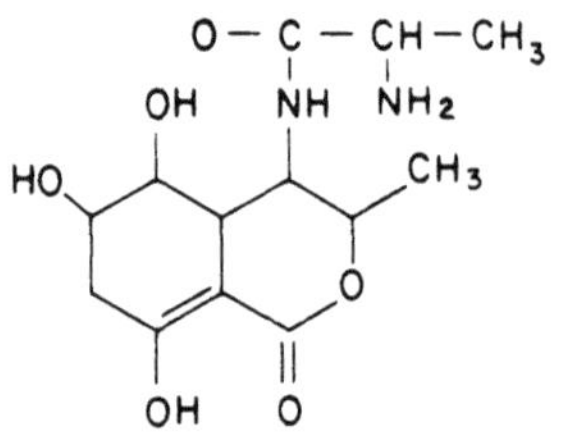

$C_{13}H_{20}O_6N_2$; MW 300.31

Fig. 39

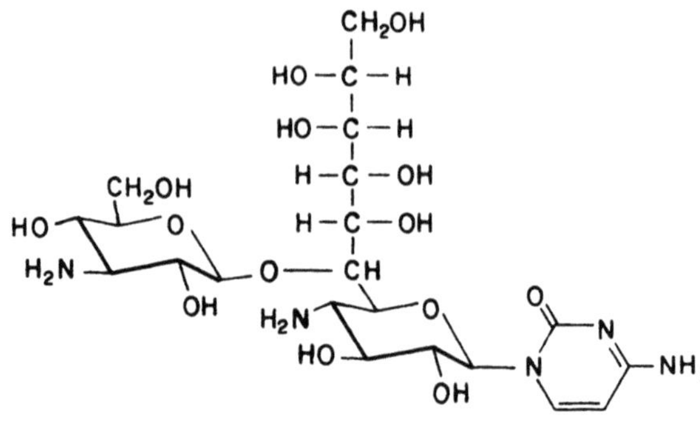

$C_{21}H_{37}O_{14}N_5$; MW 583.55

Fig. 40

Fig. 39. Chemical structure of actinobolin

Fig. 40. Chemical structure of anthelmycin

Anthelmycin (synonym hikizimycin) is an antibiotic of basic nature produced by Streptomyces longissimus (Hamill and Hoehn, 1964) and Streptomyces hikiziensis (Uchida et al., 1971). It is a broad spectrum antibiotic that is active on prokaryotic and eukaryotic organisms, mainly on ascaris, fungi, and mycobacteria. Its chemical structure is now established (Fig. 40) (Vuilhorgne et al., 1977).

Sparsomycin is produced by a strain of Streptomyces sparsogenes var. sparsogenes (Owen, Dietz and Comiener, 1962) and the chemical structure has now been established (Wiley and MacKellar, 1970) (Fig. 41). It is a broad-spectrum antibiotic

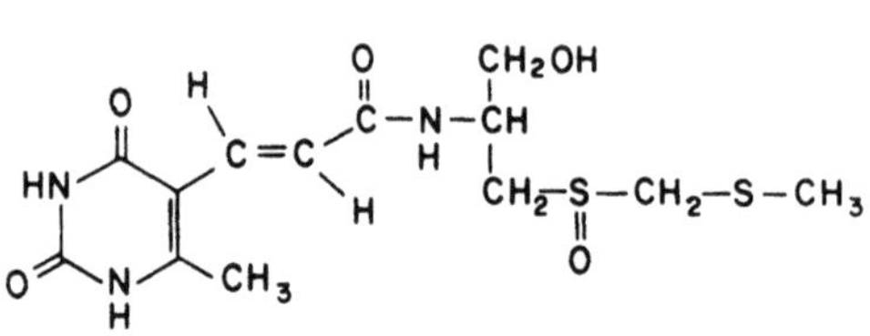

$C_{13}H_{19}O_5N_3S_2$; MW 361.43

Fig. 41. Chemical structure of sparsomycin

active on bacteria, numerous eukaryotic organisms, and as an antitumor agent. Furthermore it is widely used in biochemical studies as an inhibitor of peptide bond formation in prokaryotic and eukaryotic systems.

With the exception of puromycin, the peptidyl transferase inhibitors mentioned above are antibacterial, whereas amicetin (Battaner and Vázquez, 1971; Bloch and Coutsogeorgopoulos, 1966; Coutsogeorgopoulos, 1967; Vázquez and Monro, 1967; Carrasco and Vázquez, 1972), gougerotin (Vázquez and Monro, 1967; Clark and Gunther, 1963; Clark and Chang, 1965; Casjens and Morris, 1965; Sinohara and Sky-Peck, 1965), blasticidin S (Battaner and Vázquez, 1971; Huang, Misato and Asuyama, 1964; Yamaguchi, Yamamoto and Tanaka, 1965; Sakai and Takebe, 1970), anthelmycin (Uchida et al., 1971; Uchida and Woll, 1974), actinobolin (Battaner and Vázquez, 1971; Carrasco and Vázquez, 1972; Hunt et al., 1966; Smithers, Bennet and Struck, 1969), and sparsomycin (Colombo, Felicetti and Baglioni, 1966; Goldberg and Mitsugi, 1966, 1967a, 1967b) act on ribosomes of the prokaryotic and eukaryotic types and are mutually exclusive in their binding sites to the ribosome.

Sparsomycin is the best known antibiotic of this group. It acts on the ribosome (Trakatellis, Montjar and Axelrod, 1965) in the larger ribosomal subunit (Monro and Vázquez, 1967) and has no inhibitory effect on chloramphenicol binding to free ribosomes (Fernández-Muñoz et al., 1971; Chang, Siddhikol and Weisblum, 1969). Sparsomycin is a wide spectrum inhibitor of peptide bond formation since it inhibits the puromycin reaction in bacterial (Goldberg and Mitsugi, 1967a; Jayaraman and Goldberg, 1968; Lucas-Lenard and Haenni, 1968; Ono et al., 1969), yeast (Battaner and Vázquez, 1971), mammalian (Carrasco and Vázquez,

1972) and mitochondrial ribosomes (Sala and Kuntzel, 1970). Furthermore, sparsomycin is a powerful inhibitor of the fragment reaction assay by bacterial (Monro and Vázquez, 1967), yeast (Battaner and Vázquez, 1971), and mammalian (Carrasco and Vázquez, 1972; Neth et al., 1970) ribosomes. Sparsomycin appears to act on the acceptor site of the peptidyl transferase center since, to a certain extent, it inhibits EF-Tu-dependent binding of aminoacyl-tRNA, uncoupling GTP hydrolysis (Ravel, Shorey and Shive, 1970), and also prevents binding of CACCA-Phe (Pestka, 1969a; Celma, Monro and Vázquez, 1971; Hishizawa and Pestka, 1971), (^{3}H)puromycin (Fernández-Muñoz and Vázquez, 1973b), and (^{3}H)-gougerotin to polysomes (Barbacid and Vázquez, 1974c). Sparsomycin enhances non-enzymic binding to the donor site of Ac-Phe-tRNA (Herner, Goldberg and Cohen, 1969), even in the absence of mRNA (Jiménez, Monro and Vázquez, 1970a, 1970b), and CACCA-Leu-Ac (Monro, Celma and Vázquez, 1969). A cooperative effect on substrate- and antibiotic binding to the A- and P-sites has been repeatedly observed. Thus peptidyl-tRNA bound to the P-site enhances chloramphenicol and thiamphenicol binding (Contreras and Vázquez, 1977a) and, as indicated above, sparsomycin enhances substrate binding to the P-site. Conversely peptidyl-tRNA binding to the P-site appears to enhance sparsomycin binding to the A-site, since this antibiotic inhibits chloramphenicol binding to polysomes but not to ribosomes (Pestka, 1974). On the other hand it has been observed that chloramphenicol inhibits the sparsomycin-induced CACCA-Leu-Ac binding to ribosomes (Monro, Celma and Vázquez, 1969). The evidence reported above suggests that sparsomycin acts on the A-site of the larger subunit of prokaryotic and eukaryotic ribosomes.

Actinobolin, blasticidin S, anthelmycin, gougerotin, and the amicetin group of antibiotics (oxamicetin, bamicetin, and plicacetin) also inhibit peptide bond formation in the puromycin (Pestka, 1970a; Coutsogeorgopoulos, 1967; Jayaraman and Goldberg, 1968; Yamaguchi and Tanaka, 1966; Černá, Lichtenthaler and Rychlik, 1971) and(or) the fragment reaction assay (Battaner and Vázquez, 1971; Monro and Vázquez, 1967; Uchida and Wolf, 1974; Černá, Rychlik and Lichtenthaler, 1973; Lichtenthaler, 1973; Lichtenthaler, Černá and Rychlik, 1973, 1975). These antibiotics resemble sparsomycin in their mode of action, although they are less active. They have a certain inhibitory effect on acceptor substrate binding (Pestka, 1969a; Celma, Monro and Vázquez, 1971; Lichtenthaler, Černá and Rychlik, 1975; Hishizawa and Pestka, 1971; Černá, Lichtenthaler and Rychlik, 1971; Černá, Rychlik and Lichtenthaler, 1973), and enhance binding of the donor substrate (Celma, Monro and Vázquez, 1970; Černá and Rychlik, 1972; Černá, Lichtenthaler and Rychlik, 1971; Černá, Rychlik and Lichtenthaler, 1973; Lichtenthaler, Černá and Rychlik, 1975) to the peptidyl transferase center. Furthermore, gougerotin and blasticidin S bind to the larger ribosomal subunit (Barbacid and Vázquez, 1974c; Kinoshita, Tanaka and Umezawa, 1970). Blasticidin S binding is inhibited by gougerotin (Kinoshita, Tanaka and Umezawa, 1970). Conversely gougerotin binding to the ribosome is prevented by blasticidin S and also by actinobolin, amicetin and sparsomycin (Barbacid and Vázquez, 1974c).

Studies in cell-free systems have shown that anthelmycin very actively blocks protein synthesis by bacterial and yeast ribosomes by preventing peptide bond formation in the fragment reaction assay, and sparsomycin-induced CACCA-Leu-Ac binding to ribosomes, but it does not affect either chloramphenicol or erythromycin binding to E. coli ribosomes (González, Vázquez and Jiménez, 1978; Uchida and Wolf, 1974).

3.8 Kirromycin, Goldinomycin and Efrotomycin

Kirromycin is an antibacterial antibiotic isolated from Streptomyces collinus. It has an unusual inhibitory spectrum, being very active on some gram-positive, gram-negative and mycobacteria species but not on others (Wolf and Zähner, 1972). The antibiotics kirromycin and mocimycin appear to have an identical chemical structure (Fig. 42), whereas the antibiotic goldinomycin (synonym X-5108) is their pyridone N-methylated derivative (Vos and Verwiel, 1973; Maehr et al., 1973; Berger et al., 1973). Also closely related is the antibiotic efrotomycin which is produced by Streptomyces lactamdurans and appears to be usually more active than the others in intact bacteria, probably owing to better permeability (Wax et al., 1976).

Kirromycin has been shown to inhibit poly(U)-directed polyphenylalanine synthesis (Wolf, Zähner and Nierhaus, 1972). The

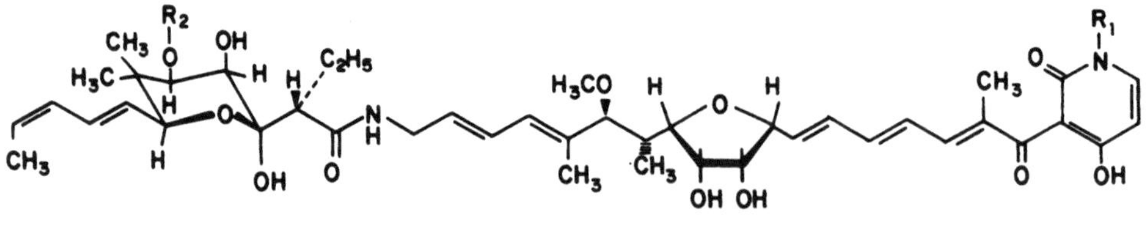

Kirromycin : R_1 = -H ; R_2 = -H ($C_{43}H_{60}O_{12}N_2$; MW 797)
Goldinomycin : R_1 = $-CH_3$; R_2 = -H ($C_{44}H_{62}O_{12}N_2$; MW 811)
Efrotomycin : R_1 = $-CH_3$; R_2 = the Disaccharide $C_{15}H_{27}O_8$ ($C_{59}H_{89}O_{20}N_2$; MW 1146)

Fig. 42. The kirromycin group of antibiotics

antibiotic specifically interacts with EF-Tu, inducing an uncoupled hydrolysis of GTP in the absence of ribosomes and aminoacyl-tRNA. However, kirromycin does not inhibit aminoacyl-tRNA binding to ribosomes (Wolf, Chinali and Parmeggiani, 1974).

Kirromycin strongly stimulates EF-Tu affinity for GTP and consequently the GDP-GTP exchange reaction with EF-Tu·GDP. It activates the EF-Tu site for aminoacyl-tRNA, abolishing the requirement of GTP for formation of the aminoacyl-tRNA·EF-Tu complex (Chinali, Wolf and Parmeggiani, 1977). The 30S ribosomal subunit alone, but not the 50S subunit, appears to be sufficient for formation of a stable complex with EF-Tu and kirromycin (Wolf, Chinali and Parmeggiani, 1977).

Kirromycin binds to EF-Tu in a 1:1 molar ratio and competes with elongation factor EF-Ts for this binding, favoring dissociation of EF-Tu and EF-Ts, probably by inhibiting their reassociation (Chinali, Wolf and Parmeggiani, 1977). The presence of either kirromycin or EF-Ts results in a 10-fold increase in the cleavage rate of EF-Tu by trypsin. Furthermore, both kirromycin and EF-Ts facilitate the exchange of free GDP with GDP bound to EF-Tu. It has therefore been suggested that kirromycin and EF-Ts might induce a similar conformational change in EF-Tu (Blumenthal, Douglass and Smith, 1977).

Kirromycin does not inhibit peptide bond formation studied in resolved systems. However, in integrated systems the interaction of kirromycin with EF-Tu results in a blockade on peptide bond formation as a consequence of the failure of EF-Tu to leave the ribosome, thus blocking interaction of the C-C-A end of aminoacyl-tRNA bound to the acceptor site with the peptidyl transferase center. In agreement with these findings there is a single round of ribosome and EF-Tu-dependent GTP hydrolysis in the presence of poly (U), Phe-tRNA^{Phe} and kirromycin (Wolf, Chinali and Parmeggiani, 1977).

Mutants of E. coli resistant to kirromycin have been described in which resistance is due to an alteration of EF-Tu. Kirromycin-resistant EF-Tu is coded by an altered tuf B_1 gene. Kirromycin sensitivity is dominant as shown by a failure of a mixture of EF-Tu population to express resistance in vitro. Since there are two genes coding for EF-Tu (tuf A and tuf B_1), phenotypic expression of kirromycin resistance in vivo appears to be only possible if the EF-Tu mutant lacks an active tuf A gene (Fischer et al., 1977; Van de Klundert et al., 1977).

3.9 Anisomycin

The antibiotic anisomycin was first isolated by Sobin and Tanner (1954) from culture filtrates of two different species of Streptomyces, subsequently identified as S. griseolus and S. roseochromogenes (Tanner, Sobin and Gardocki, 1955). The antibiotic is mainly active on protozoa but it is also an inhibitor of yeast, fungi, and other eukaryotic cells. Both the chemical structure and the stereochemical configuration of the drug were established and its complete chemical synthesis has also been achieved (Jiménez and Vázquez, 1978, review) (Fig. 43).

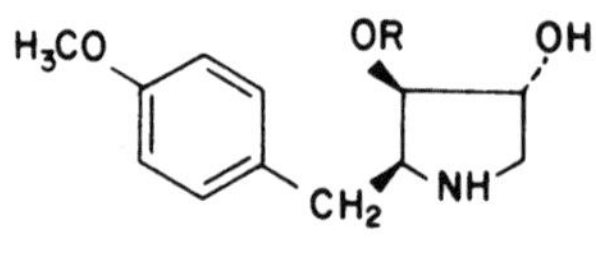

Anisomycin : R = - CO — CH_3 (C_{14} H_{19} O_4 N ; MW 265.30)

Deacetylanisomycin : R = - H

Fig. 43. Chemical structure of anisomycin

Anisomycin inhibits protein synthesis in eukaryotic systems (Grollman, 1967a; Bermek and Matthaei, 1970) by specifically blocking peptide bond formation as shown in the fragment reaction (Vázquez et al., 1969; Battaner and Vázquez, 1971; Carrasco and Vázquez, 1972; Neth et al., 1970) and in the puromycin reaction (Carrasco and Vázquez, 1972; Grollman, 1967a). Competitive inhibition of puromycin action by anisomycin has been reported (Pestka et al., 1972). To a certain extent anisomycin inhibits substrate interaction with the donor and acceptor sites of the peptidyl transferase center (Battaner and Vázquez, 1971).

Anisomycin specifically binds to the 60S ribosomal subunit; there is a single binding site for the antibiotic per ribosome (Barbacid and Vázquez, 1973; Barbacid and Vázquez, 1975), this binding being inhibited by tenuazonic acid and compounds of the trichodermin and narciclasine (Barbacid and Vázquez, 1974a) groups as well as by rather high concentrations of streptogramin A and griseoviridin (Barbacid, Contreras and Vázquez, 1975; Barbacid and Vázquez, 1974a). Conversely anisomycin prevents (^{14}C)trichodermin (Barbacid and Vázquez, 1974b) and (^{3}H)-narciclasine binding (Baez and Vázquez, 1978) with the ribosome. A mutation conferring resistance to anisomycin and trichodermin is expressed in the 60S subunit of the yeast ribosome (Jiménez, Sánchez and Vázquez, 1975a; Jiménez and Vázquez, 1975). Both sensitive and resistant ribosomes bind (^{3}H)anisomycin equally well but (^{14}C)trichodermin binding by mutant ribosomes is less efficient than in wild type ribosomes (Jiménez, Sánchez and Vázquez, 1975a; Jiménez and Vázquez, 1975). These results suggest that trichodermin and anisomycin might not act at precisely the same ribosomal site. Heterogeneous binding of (^{3}H)anisomycin to polysomes can be observed, since the interaction of the antibiotic is considerably enhanced when peptidyl-tRNA is bound to the P-site (Barbacid and Vázquez, 1975).

There are a number of compounds whose chemical structure is closely related to that of anisomycin and which are also active in eukaryotic cells (Ishida et al., 1974); however, most of them were found not to be inhibitors of translation (Jiménez and Vázquez, 1978). Particularly interesting is deacetylanisomycin (Fig. 43), which interacts with the ribosome at the same site as anisomycin but has a much lower affinity (Barbacid and Vázquez, 1975).

There is an increasing number of peptide bond formation inhibitors acting on eukaryotic ribosomes at sites overlapping or closely connected with that of anisomycin. Their effects are briefly summarized in Table 10.

3.10 Trichothecene Antibiotics

Trichothecene antibiotics are compounds of sesquiterpene nature (12-13 epoxytrichothecene) whose chemical structures have been established (Fig. 44) and which were recently reviewed (Bamburg, 1976; Ueno, 1977). There are more than 30 compounds within this group; the best known of them is trichodermin, which is produced by *Trichoderma viride* cultures (Godfredsen and Vangedal, 1964), while other antibiotics in the group are produced by different fungi (Table 11). Their mode of action has been reviewed recently (Jiménez and Vázquez, 1978). Trichothecene antibiotics are very effective antifungal agents and, in general, are also active on different eukaryotic cells.

a

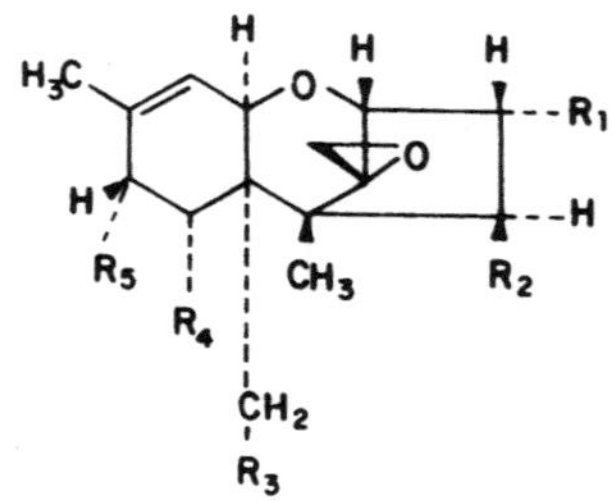

	R_1	R_2	R_3	R_4	R_5	
Scirpene	H	H	H	H	H	
Trichodermol	H	OH	H	H	H	($C_{15}H_{21}O_3$; MW 249.32)
Trichodermin	H	$O\text{-}OC\text{-}CH_3$	H	H	H	($C_{17}H_{24}O_4$; MW 292.36)
Diacetoxyscirpenol	OH	$O\text{-}OC\text{-}CH_3$	$O\text{-}OC\text{-}CH_3$	H	H	($C_{19}H_{26}O_7$; MW 366.40)
Toxin T-2	OH	$O\text{-}OC\text{-}CH_3$	$O\text{-}OC\text{-}CH_3$	H	$OOCCH_2\text{-}CH(CH_3)_2$	($C_{24}H_{34}O_9$; MW 466.51)
Verrucarol	H	OH	OH	H	H	
15-Deacetylcalonectrin	$O\text{-}OC\text{-}CH_3$	H	OH	H	H	
Scirpenol	OH	OH	OH	H	H	

b

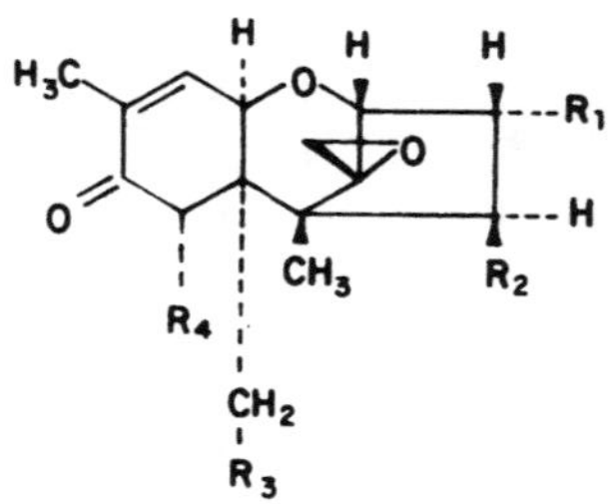

	R_1	R_2	R_3	R_4	
Nivalenol	OH	OH	OH	OH	($C_{15}H_{20}O_7$; MW 312.31)
Trichothecin	H	$OOCCH{=}CH\text{-}CH_3$	H	H	($C_{19}H_{24}O_5$; MW 332.37)
Fusarenon X	OH	$O\text{-}OC\text{-}CH_3$	OH	OH	($C_{17}H_{22}O_8$; MW 354.35)

c

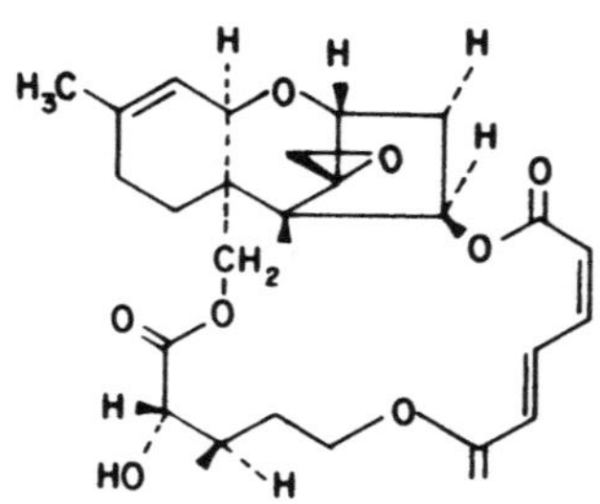

Verrucarin A ($C_{27}H_{34}O_9$; MW 502.54)

Fig. 44. Trichothecene antibiotics

Table 10. Comparative effects of anisomycin and related compounds on different reactions

Compound	Inhibition of initiation	Inhibition of peptide bond formation by sensitive ribosomes	Inhibition of peptide bond formation by trichodermin-resistant ribosomes from yeast
Anisomycin	-	+	-
Trichodermin	-	+	-
Trichothecin	-	+	-
Fusarenon X	-	+	-
Verrucarin A	-	+	-
Tenuazonic acid	-	+	a
Narciclasine	-	+	-
Pretazettine	-	+	-
Haemanthamine	-	+	-
Lycorine	-	+	+
Pseudolycorine	-	+	+
Harringtonine	-	+	-
Homoharringtonine	-	+	-
Isoharringtonine	-	+	-
Bruceantin	-	+	-

[a]Tenuazonic acid is not active on yeast ribosomes even from the wild strain.
[b]Fusarenon X, at rather high concentrations, inhibits peptide bond formation by polysomes.

Table 10 (continued)

Inhibition of peptide bond formation by polysomes	Inhibition of (^{3}H)-anisomycin, (^{14}C)-trichodermin, and (^{3}H)narciclasine binding to ribosomes	High affinity binding to ribosomes of a trichodermin resistant strain from yeast
+	+	+
+	+	-
+	+	-
+(b)	+	-
-	+	-
+	+	a
+	+	-
+	+	-
+	+	-
+	+	++
+	+	++
-	+	-
-	+	-
-	+	-
-	+	+

Table 11. Trichothecene antibiotics: sources

Antibiotics	Producer fungi
Trichodermin	Trichoderma viride
Trichodermol (synonym roridin C)	Myrothecium roridum
Verrucarin A	Myrothecium verrucaria Myrothecium roridum
Trichothecin	Trichothecium roseum
T-2 toxin	Fusarium tricinctum
Fusarenon X	Fusarium nivale Fusarium episharia
Crotocin (synonym antibiotic-T)	Cephalosporium crotocinigenum
Calonectrin	Fusarium nivale
Triacetoxyscirpendiol	Fusarium roseum
Nivalenol	Fusarium nivale
Diacetoxyscirpenol	Fusarium solani

Trichothecene antibiotics strongly inhibit protein synthesis in eukaryotic cells and cell-free systems (Carrasco, Barbacid and Vázquez, 1973; Ueno et al., 1968; Ueno et al., 1969; Stafford and McLaughlin, 1973; Tate and Caskey, 1973). Different toxins of this group (trichodermin, trichodermol, trichothecin, verrucarin A, fusarenon X, nivalenol, diacetoxyscirpenol, and T-2 toxin) are very active in blocking the peptide bond formation

(Carrasco, Barbacid and Vázquez, 1973; Ueno et al., 1968; Ueno et al., 1969; Stafford and McLaughlin, 1973; Tate and Caskey, 1973) and termination reactions (Stafford and McLaughlin, 1973; Tate and Caskey, 1973). However, on the basis of their effects on polysomes in intact cells (Cundliffe, Cannon and Davies, 1974) or cell-free systems (Schindler, 1974; Liao, Grollman and Horwitz, 1976) these trichothecene toxins have been classified into two groups (Schindler, 1974). Thus trichodermin, trichodermol, crotocin and trichothecin freeze the polysomes and prevent the puromycin reaction with polysomes and have therefore been classified as inhibitors of peptide bond formation (Cundliffe, Cannon and Davies, 1974; Schindler, 1974; Liao, Grollman and Horwitz, 1976). On the other hand verrucarin A, nivalenol, T-2 toxin, and diacetoxyscirpenol cause polysome breakdown and do not prevent the puromycin reaction with polysomes and on these grounds have been considered inhibitors of initiation (Cundliffe, Cannon and Davies, 1974; Schindler, 1974; Liao, Grollman and Horwith, 1976; Mizuno, 1975; Cundliffe and Davies, 1977). However, it is now clear that none of the trichothecene antibiotics inhibits initiation but all of them are inhibitors of peptide bond formation (Carrasco, Barbacid and Vázquez, 1973; Smith, Cannon and Cundliffe, 1975; Cannon, Jiménez and Vázquez, 1976). Nevertheless, trichodermin, trichodermol, crotocin, and trichothecin interact with polysomes as well as ribosomes, whereas the affinity of verrucarin A, nivalenol, T-2 toxin, and diacetoxyscirpenol for the polysomes is so low that they only inhibit the first few steps of peptide bond synthesis prior to polysome formation (Cannon, Jiménez and Vázquez, 1976; Cannon, Smith and Carter, 1976; Carter, Cannon and Smith, 1976). An intermediate situation occurs with fusarenon

X since at low concentrations it inhibits only the first few steps of peptide bond formation and therefore causes polysome run-off, whereas at higher concentration it can act on polysomes preventing their breakdown (Carter and Cannon, 1977, 1978).

All trichothecene antibiotics which have been tested, block strongly the interaction with free ribosomes of peptide bond formation inhibitors such as (^{3}H)anisomycin (Barbacid and Vázquez, 1974a) and (^{3}H)narciclasine (Baez and Vázquez, 1978). One molecule of (^{14}C)trichodermin binds to sensitive ribosomes (Barbacid and Vázquez, 1974b; Wei et al., 1974) on their 60S subunits but not to 40S subunits (Barbacid and Vázquez, 1974b). Binding of the antibiotic to washed ribosomes is prevented by the different active toxins of the trichothecene group, anisomycin and narciclasine (Barbacid and Vázquez, 1974b; Wei et al., 1974). However, (^{14}C)trichodermin binding to polysomes is not affected by those toxins indicated above that do not freeze polysomes (Cannon, Jiménez and Vázquez, 1976; Vázquez et al., 1977). The isolation of mutants resistant to trichodermin that display cross-resistance to other trichothecene toxins, anisomycin (Schindler, Grant and Davies, 1974; Jiménez, Sánchez and Vázquez, 1975a; Jiménez and Vázquez, 1975), and narciclasine (Jiménez, Sánchez and Vázquez, 1975b, 1975c) further suggests that all these inhibitors have intimately related binding sites on the 60S ribosomal subunit. Resistance to trichodermin is due to a single recessive gene and the mutation was found to be located on the right arm of chromosome XV in _Saccharomyces cerevisiae_ (Grant, Shindler and Davies, 1976).

3.11 Tenuazonic Acid

Tenuazonic acid is an antibiotic of simple chemical structure (Fig. 45) which was initially obtained from cultures of *Alternaria tenuis* (Rosett et al., 1957). The antibiotic is only active on some eukaryotic cells and as an antitumor agent.

Tenuazonic acid inhibits mammalian protein synthesis (Shigeura and Gordon, 1963) by preventing peptide bond formation (Carrasco and Vâzquez, 1972, 1973b; Neth et al., 1970). Tenuazonic acid acts on the 60S ribosomal subunit (Carrasco and Vâzquez, 1973b) and prevents (^{3}H)anisomycin (Barbacid and Vâzquez, 1974a) and (^{14}C)trichodermin (Barbacid and Vâzquez, 1974b) interaction. An important characteristic of tenuazonic acid is its preferential inhibition of mammalian systems; studies on inhibition of protein synthesis (Carrasco and Vâzquez, 1973b) and (^{3}H)anisomycin binding (Barbacid and Vâzquez, 1974a) have shown the antibiotic to have less effect on plant ribosomes, and to be inactive on yeast ribosomes.

3.12 Narciclasine and Related Compounds

Narciclasine is a compound which accumulates in the bulbs of different species of Narcissus and has antitumor activity blocking the growth of eukaryotic cells (Ceriotti, 1967) (Fig. 46). The

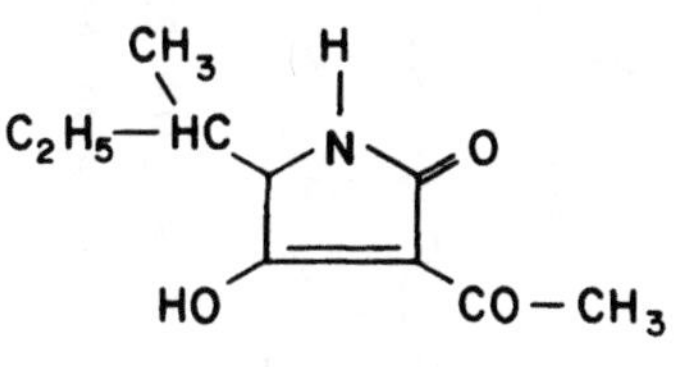

$C_{10}H_{15}O_3N$; MW 197.23

Fig. 45. Chemical structure of tenuazonic acid

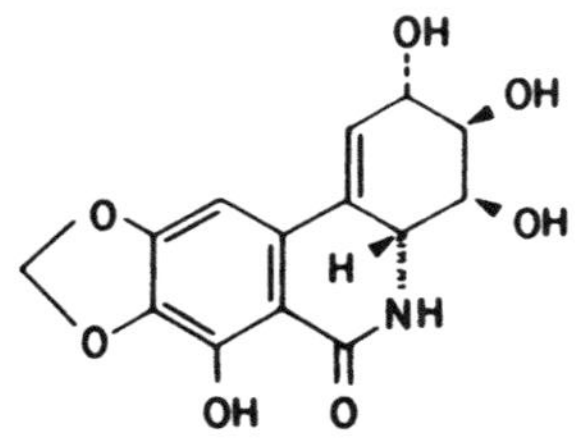

$C_{14}H_{13}O_7N$; MW 307.25
Narciclasine

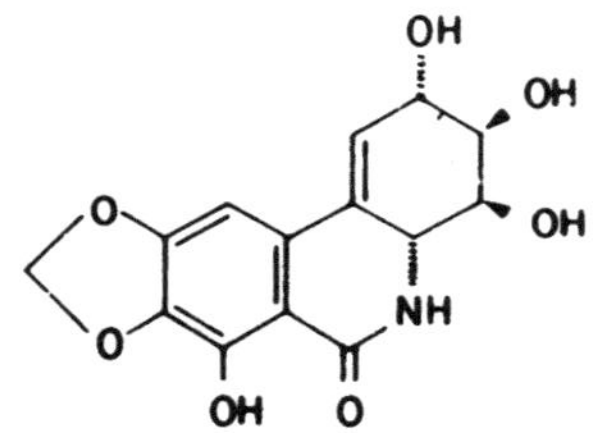

Isonarciclasine

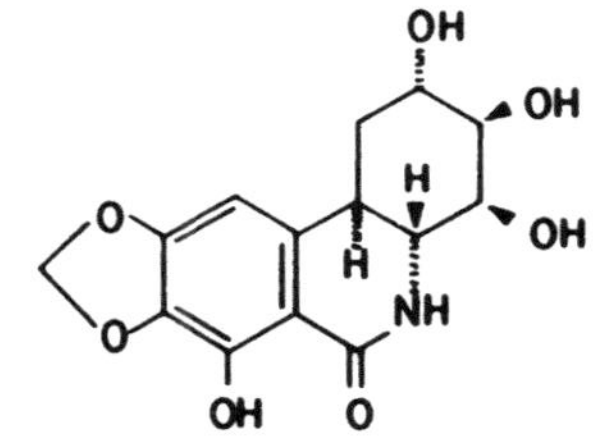

Trans - dihydronarciclasine

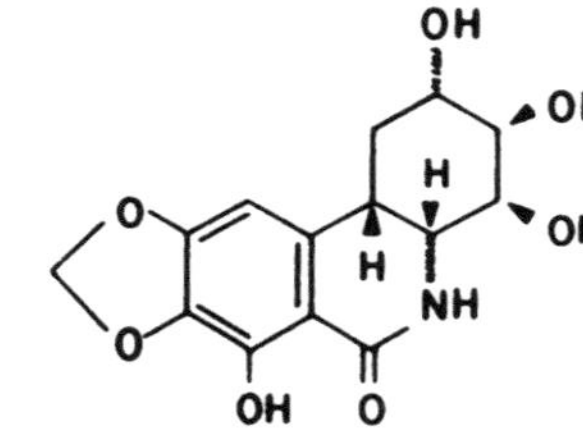

Cis - dihydronarciclasine

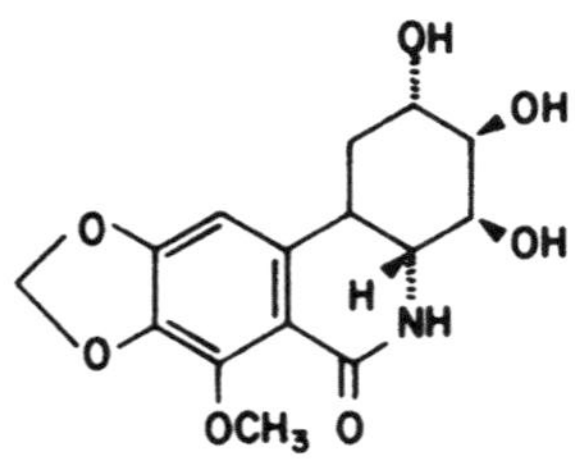

O - Methylnarciclasine

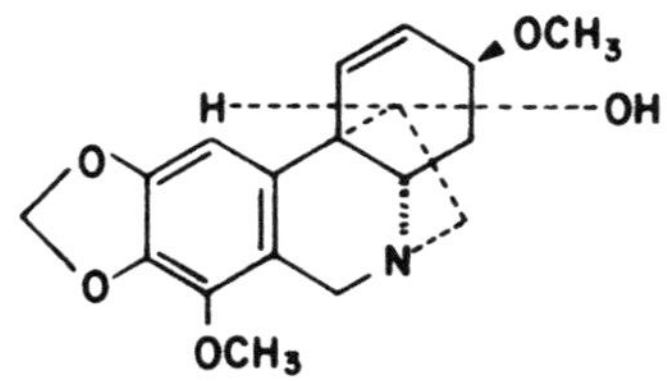

$C_{17}H_{19}O_4N$; MW 301.33
Haemanthamine

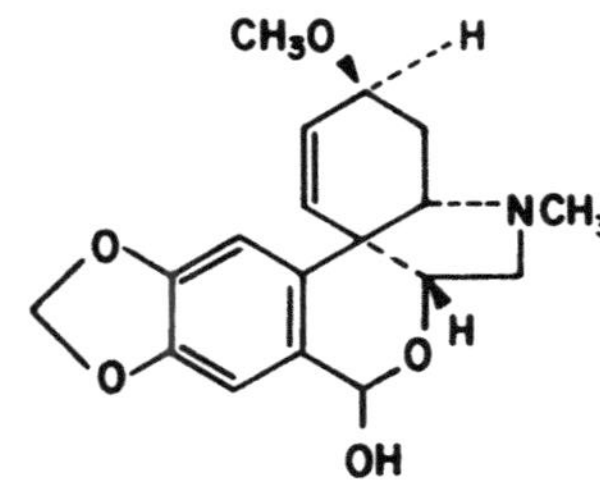

$C_{18}H_{19}O_5N$; MW 339.34
Pretazettine

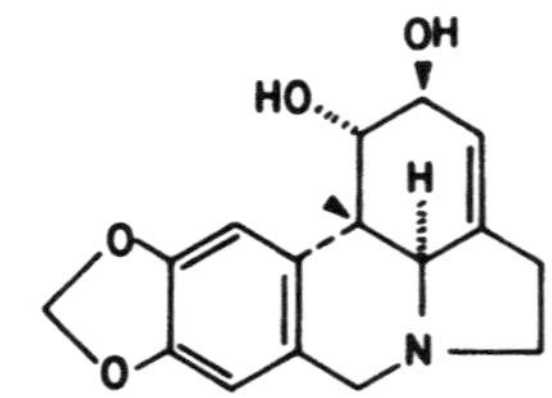

$C_{16}H_{17}O_4N$; MW 287.30
Lycorine

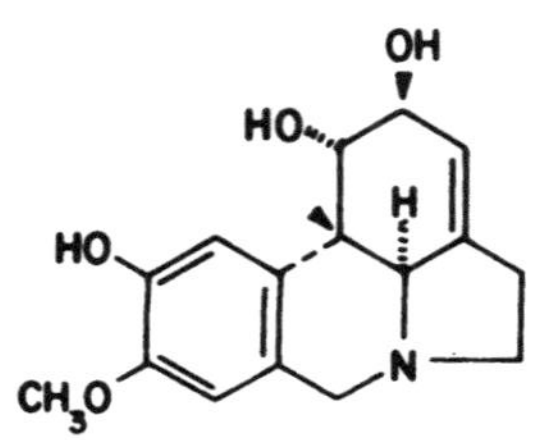

$C_{16}H_{19}O_4N$; MW 289.32
Pseudolycorine

Fig. 46. Chemical structures of narciclasine and related compounds

mode of action of narciclasine and related Amaryllidaceae alkaloids has been reviewed recently (Jiménez and Vázquez, 1978). Narciclasine specifically inhibits protein synthesis in intact cells (Jiménez et al., 1976) and cell-free systems (Carrasco, Fresno and Vázquez, 1975) by preventing peptide bond formation as shown by the fragment and puromycin reaction assays with ribosomes (Carrasco, Fresno and Vázquez, 1975; Jiménez et al., 1976). Furthermore, narciclasine blocks the puromycin reaction with polysomes and the sparsomycin-induced binding of UACCA-Leu-Ac to ribosomes (Carrasco, Fresno and Vázquez, 1975). One molecule of (^{3}H)narciclasine binds per ribosome on the 60S ribosomal subunit (Baez et al., 1977). Narciclasine drastically inhibits (^{3}H)anisomycin binding and to a lesser extent (^{14}C)trichodermin binding to ribosomes (Carrasco, Fresno and Vázquez, 1975). Conversely anisomycin and trichodermin prevent (^{3}H)narciclasine interaction with the ribosomes (Baez and Vázquez, 1978). Ribosomes from a Saccharomyces cerevisiae mutant resistant to trichodermin are also resistant to narciclasine (Jiménez, Sánchez and Vázquez, 1975b, 1975c). All these results clearly show that narciclasine is an inhibitor of the peptidyl transferase center of eukaryotic ribosomes acting at a site closely connected to or overlapping the binding sites of anisomycin and trichodermin. There is an ample number of compounds whose chemical structure is closely related to that of narciclasine; some have been obtained from Narcissus bulbs and others by chemical synthesis (Mondon and Krohn, 1975). A number of these compounds are also inhibitors of protein synthesis and prevent (^{3}H)-narciclasine binding to ribosomes (isonarciclasine, trans- and cis-dihydronarciclasine and O-methylnarciclasine) (Fig. 46), whereas others are devoid of biological activity (Baez and

Vázquez, 1978). In the bulbs of Amaryllidaceae there are more than 100 alkaloids in some way related to narciclasine in their chemical structure (Wildman, 1968; Fuganti, 1975). A number of these alkaloids, including pretazettine, haemanthamine, lycorine and pseudolycorine (Fig. 46), also inhibit protein synthesis by blocking peptide bond formation (Jiménez et al., 1976) and prevent (^{3}H)narciclasine binding to ribosomes (Baez and Vázquez, 1978). Ribosomes from a Saccharomyces cerevisiae mutant resistant to trichodermin are also resistant to narciclasine, pretazettine, and haemanthamine but not to lycorine and pseudolycorine, which suggests that these last two alkaloids do not act on the peptidyl transferase center on precisely the same site as narciclasine (Jiménez, Sánchez and Vázquez, 1975b). Although pretazettine and narciclasine appear to have the same site of action on the peptidyl transferase center, pretazettine demonstrates remarkable activity against lymphocytic choriomeningitis virus infections in mice (Furusawa and Cutting, 1970) and rauscher leukemia (Furusawa et al., 1976) while narciclasine does not. Furthermore pretazettine, at rather high concentrations, also has a certain inhibitory effect on reverse transcriptase (Papas, Sandhaus and Chirigos, 1973). The effects of these compounds and other closely related peptide bond formation inhibitors are summarized in Table 10.

3.13 Cephalotaxus Alkaloids

The alkaloids harringtonine, homoharringtonine and isoharringtonine are obtained from seeds of Cephalotaxus harringtonia. Their chemical structure has already been established (Fig. 47); they are provided with antitumor activity (Powell, Weisleder and Smith,

1972) since they are very effective inhibitors of protein synthesis in eukaryotic cells and cell-free systems (Tscherne and Pestka, 1975; Huang, 1976), therefore blocking the cell cycle in the G_1 and G_2 phases (Baaske and Heinstein, 1977). These alkaloids were initially proposed as inhibitors of initiation since they cause polysome breakdown in intact cells and cell-free systems (Tscherne and Pestka, 1975; Huang, 1976). However, this effect is due to the very low affinity of the harringtonine alkaloids for polysomes (Fresno, Jiménez and Vázquez, 1977). Indeed these compounds do not inhibit any of the initiation steps with either derived 40S (Fresno, Jiménez and Vázquez, 1977) or native 40S ribosomal subunits (Fresno and Vázquez, 1978). On the other hand the <u>Cephalotaxus</u> alkaloids not only inhibit EF-1-dependent binding of aminoacyl-tRNA to ribosomes and peptide bond formation by ribosomes, but also prevent the interaction of other peptide bond formation inhibitors such as (^{14}C)trichodermin, (^{3}H)anisomycin (Fresno, Jiménez and Vázquez, 1977), and (^{3}H)narciclasine (Baez and

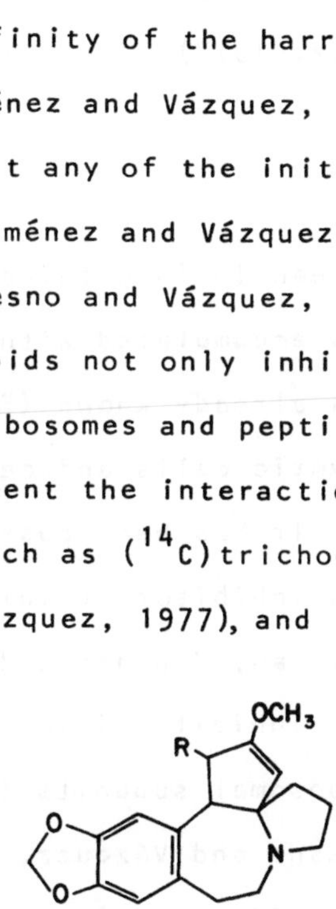

Harringtonine ($C_{28}H_{37}O_9N$; MW 531.58). R= $CH_3-C(OH)(CH_3)-CH_2-CH_2-C(OH)(O{=}CO-)-CH_2-COO-CH_3$

Isoharringtonine ($C_{29}H_{37}O_9N$; MW 543.59). R= $CH_3-CH(CH_3)-CH_2-CH_2-C(OH)(O{=}C-)-CH(OH)-COO-CH_3$

Homoharringtonine ($C_{28}H_{37}O_9N$; MW 531.58). R= $CH_3-C(OH)(CH_3)-CH_2-CH_2-CH_2-C(OH)(O{=}C-)-CH_2-COO-CH_3$

<u>Fig. 47.</u> <u>Cephalotaxus</u> alkaloids

Vázquez, 1978) with the ribosome. Nevertheless, the Cephalotaxus alkaloids have a low affinity for polysomes since they are very poor inhibitors of (^{14}C)trichodermin binding to ribosomes in polysomal preparations (Fresno, Jiménez and Vázquez, 1977). Therefore, the Cephalotaxus alkaloids block translation in the early rounds of elongation prior to polysome formation and have no effect on any of the model steps of the initiation process (Fresno, Jiménez and Vázquez, 1977).

3.14 Bruceantin

The antitumor compound bruceantin is obtained from *Brucea antidysenterica* where it is accumulated with some related compounds whose chemical structure is already known (Fig. 48). It inhibits protein synthesis in eukaryotic cells and cell-free systems (Liao, Kupchan and Horwitz, 1976). It has been postulated that bruceantin at low concentrations is an inhibitor of initiation since it causes polysome breakdown (Liao, Kupchan and Horwitz, 1976). However, bruceantin does not inhibit initiation of translation with either derived 40S ribosomal subunits (Fresno et al., 1978) or native 40S subunits (Fresno and Vázquez, 1978). On the other hand bruceantin inhibits peptide bond formation by ribosomes and

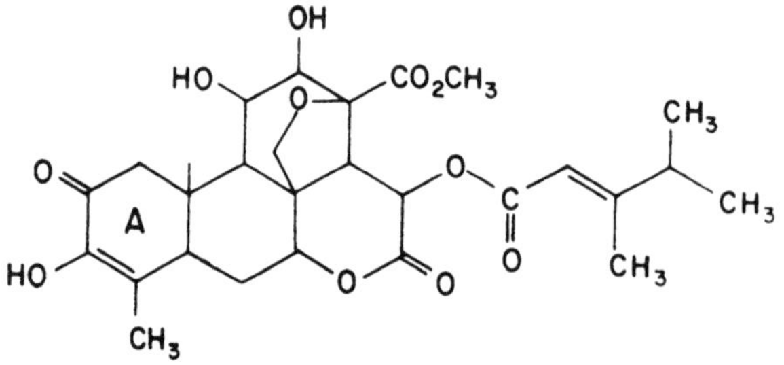

$C_{27}H_{35}O_{11}$; MW 535.51

Fig. 48. Chemical structure of bruceantin

also prevents the interaction with the ribosome of known peptide bond formation inhibitors such as (^{14}C)trichodermin (Fresno et al., 1978) and (^{3}H)narciclasine (Baez and Vázquez, 1978). Even so bruceantin has very low affinity for polysomes since it is a very poor inhibitor of (^{14}C)trichodermin binding to polysomes (Vázquez et al., 1977; Fresno et al., 1978). Indeed, bruceantin inhibits only the first rounds of peptide bond formation prior to polysome formation (Fresno et al., 1978).

3.15 Thiopseudourea

S-(10-undecen-1-yl)isothiouronium iodide (AHR-1911) is a synthetic compound with bactericide and fungicide action. The compound also has antitumor activity and interferes with nucleic acid and protein synthesis in liver. In cell-free synthesis it inhibits protein synthesis by cytoplasmic ribosomes and to a lesser extent by mitochondrial ribosomes. Thiopseudourea was proposed as a specific inhibitor of peptide bond formation since it blocks the puromycin reaction very effectively (González-Cadavid and Herrera Quijada, 1974).

4. Inhibitors of Translocation [Step (f)]

4.1 Diphtheria Toxin

Biochemical studies on diphtheria toxin have recently been reviewed (Pappenheimer, 1977). Early studies clearly showed that although the bacteria *Corynebacterium diphtheriae* is located in the throat of diphtheria patients, other lesions are produced by an extracellular toxin, the diphtheria toxin, which is produced by the bacteria and transported by the blood to other tissues. However only lysogenic strains of C. diphtheriae infected with bacteriophage β (C. diphtheriae $(\beta)^{tox\ +}$) can cause diphtheria, since the toxin is coded by the tox gene of the genome of bacteriophage β; thus non lysogenic strains of C. diphtheria not infected with the bacteriophage can develop in the throat,causing some fever but not producing either the toxin or the diphtheria (Pappenheimer and Gill, 1973). On the other hand the diphtheria toxin in the C. diphtheriae $(\beta)^{tox\ +}$ can account for over 5% of the total protein produced by the bacteria.

Diphtheria toxin is a protein (MW = 62,000) which upon reduction of a disulphide bridge and hydrolysis of a peptide bond, gives two polypeptides (MW = 24,000 and MW = 38,000) (Fig. 49). This hydrolysis can be carried out in cell-free systems, but also takes place when the diphtheria toxin interacts with the eukaryotic cell. The larger polypeptide fragment is involved in the penetration of the toxin into the cell whereas the smaller one (fragment A) inhibits protein synthesis (Pappenheimer and Gill, 1972, 1973; Collier, 1975; reviews). The amino acid sequence of the enzymically active fragment A has been elucidated (Delunge, Drazin and Collier, 1976). Diphtheria toxin also inhibits amino acid incorporation by eukaryotic cell-free systems (Pappenheimer and Gill, 1972, 1973; reviews). NAD is absolutely necessary for the inhibitory action of the toxin (Collier and Pappenheimer, 1964), which takes place by inactivation of EF-2 (Collier, 1967; Goor and Pappenheimer, 1967) and therefore prevents translocation (Carrasco and Vázquez, 1972; Schneider, Raeburn and Maxwell, 1968; Tiboni and Ciferri, 1971). The enzymic inactivation of EF-2 by diphtheria toxin takes place according to the reaction:

$EF\text{-}2 + NAD^+ \rightleftharpoons$ ADP-ribosyl-EF-2 + nicotinamide + H^+
(Honjo, Nishizuka and Hayaishi, 1969; Gill, Pappenheimer and Baseman, 1969; Everse et al., 1970)

The ADP-ribosyl- is linked to an unknown and weakly basic α amino acid residue present in EF-2 from rat liver and rabbit reticulocytes to the extent of only one residue per mole (Robinson and Maxwell, 1972; Maxwell, Robinson and Henriksen,

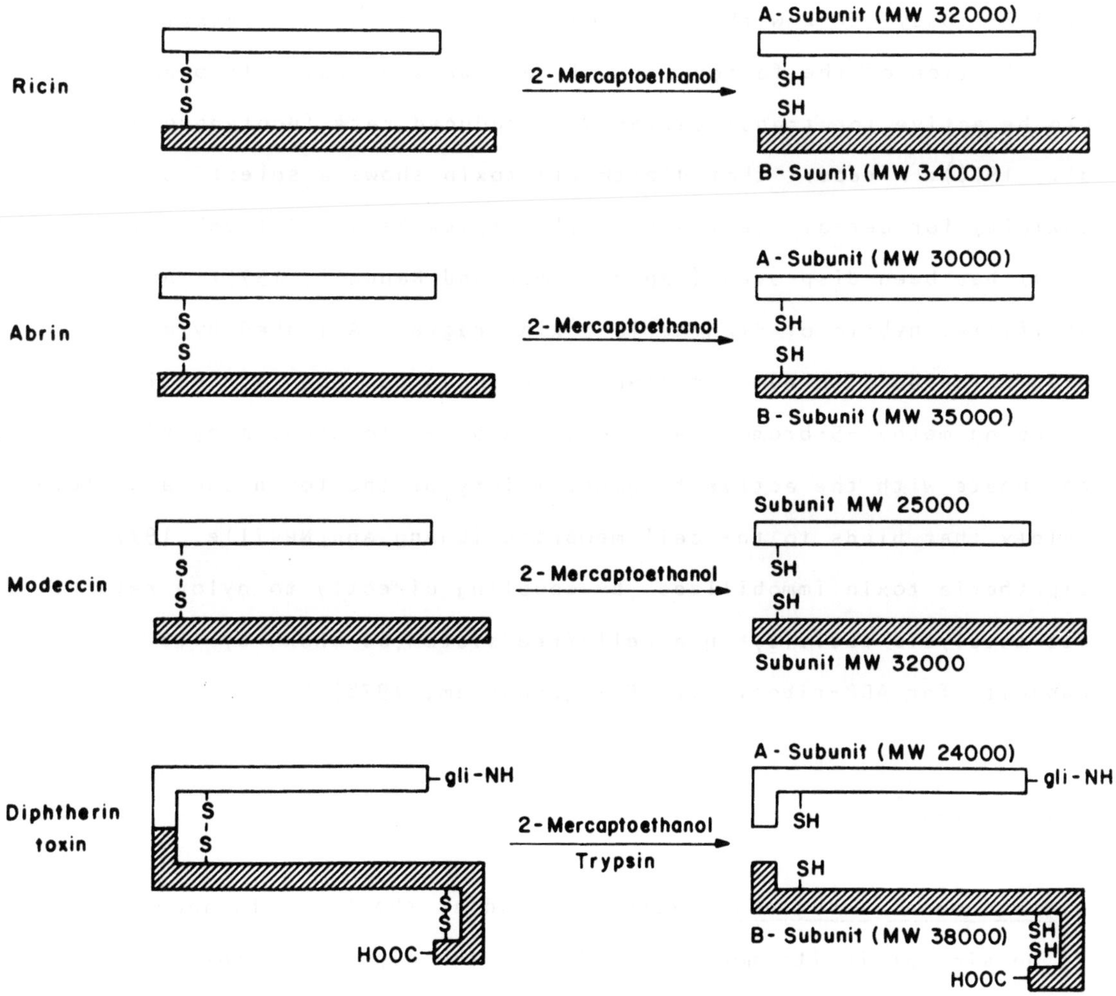

Fig. 49. Structures of some toxins

1975). The ADP-ribosyl-EF-2-GTP-ribosome complex is formed (Bermek, 1972) and, in fact, ADP-ribosyl-EF-2 competes with EF-2 for the formation of this complex with GTP and the ribosome (Bermek, 1976). Hydrolysis of the GTP in the complex also takes place (Tiboni and Ciferri, 1971). ADP-ribosyl-EF-2 is partially active for translocation although its efficiency is rather low but it is practically inactive in complementing EF-1 for polypeptide synthesis (Montanaro et al., 1976). It appears that a concomitant decrease in the rate of association and dissociation of ADP-ribosyl-EF-2 with the ribosome results in this drastic inactivation of the factor for polypeptide synthesis,although it can be active for translocation at a reduced rate (Montanaro et al., 1976). A report that diphtheria toxin shows a selective toxicity for certain malignant cells (Iglewski and Rittenberg, 1974) has been disproved (Pappenheimer and Randall, 1975). An artificial hybrid of diphtheria toxin fragment A linked by a disulphide bond to human placental lactogen has been synthesized by using methyl-5-bromovalerimidate in order to study a hybrid conjugate with the active fragment moiety of the toxin and a protein moiety that binds to the cell membrane (Chang and Neville, 1977). Diphtheria toxin immobilized by coupling directly to nylon retains its catalytic activity in a cell-free system,as shown by its capacity for ADP-ribosylate EF-2 (Sundaram, 1978).

4.2 PA Toxin

A *Pseudomonas aeruginosa* exotoxin known as the PA toxin appears to be similar in its mode of action to the diphtheria toxin (Iglewski and Kabat, 1975). However no subunits have been separated in the PA toxin (MW = 50,000 - 54,000) that have molecular and immunological properties different from those of the diphtheria toxin (Iglewski and Kabat, 1975). In a recent work MW of the toxin was found to be 66,000 and a fragment of 26,000 dalton was observed to be enzymically active in cell-free systems (S. Lory and R.J. Collier, personal communication).

4.3 Cycloheximide and Related Glutarimide Antibiotics

The antibiotic cycloheximide (synonyms actidione and naramycin A) was first isolated from culture filtrates from *Streptomyces griseus* in mixtures with streptomycin (Whiffen, Bohonos and Emerson, 1946). The antibiotic is also produced by a number of other *Streptomyces* species including *S. noursei* and *S. naraensis*. The chemical structure of cycloheximide was early established (Kornfeld and Jones, 1948) and confirmed by synthesis (Johnson et al., 1964). A number of antibiotics have been isolated from different *Streptomyces* species including inactone, streptimidone, streptovitacin A,and actiphenol which, like cycloheximide, have a glutarimide structure (Fig. 50) and are inhibitors of protein synthesis (Korzybski, Kowszyk-Gindifer and Kurylowicz, 1967a,b; Sisler and Siegel, 1967; reviews).

Cycloheximide was early described as an inhibitor of protein synthesis in eukaryotic cells (Kerridge, 1958) and cell-free systems (Felicetti, Colombo and Baglioni, 1966; Siegel and Sisler, 1963; Ennis and Lubin, 1964; Bennett, Ward and Brockman, 1965; Viau and Davis, 1970). Apart from protein synthesis, cycloheximide also inhibits DNA and RNA synthesis in intact cells (Sisler and Siegel, 1967; Viau and Davis, 1970; Timberlake and

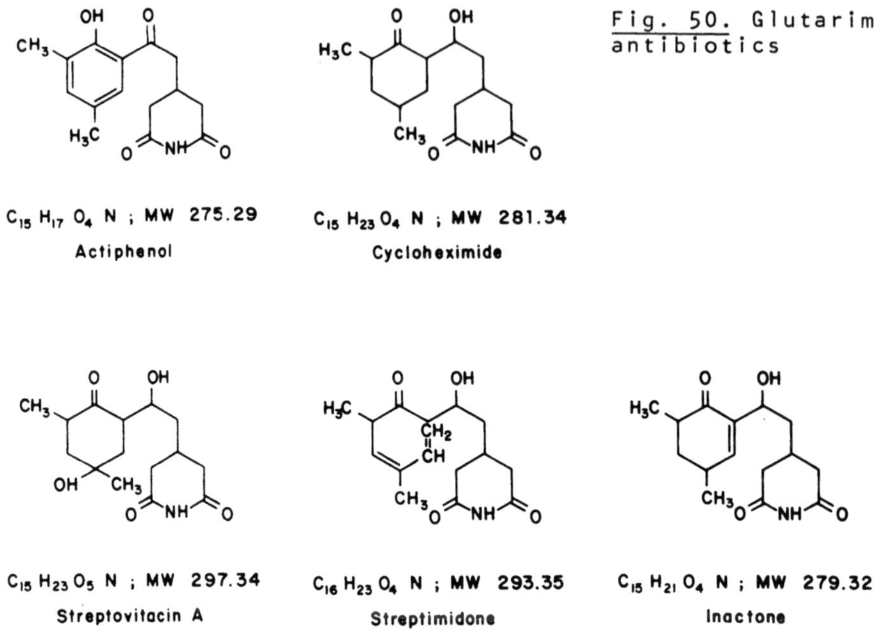

Fig. 50. Glutarimide antibiotics

Griffin, 1974). Furthermore, inhibition by cycloheximide of RNA polymerase I from some fungi has been reported (Horgen and Griffin, 1971; Timberlake, McDowell and Griffin, 1972). Although it was initially thought that cycloheximide interacted with the elongation factors (Felicetti, Colombo and Baglioni, 1966; Trakatellis, Montjar and Axelrod, 1965), the ribosome was later suggested as the site of interaction of the antibiotic (Battaner and Vâzquez, 1971). This is now widely accepted after experiments using ribosomes from cycloheximide-resistant strains (Siegel and Sisler, 1965; Cooper, Banthorpe and Wilkie, 1967; Rao and Grollman, 1967; Jiménez, Littlewood and Davies, 1972; Pongratz and Klingmuller, 1973; Poche et al., 1975). It was further concluded that the 60S subunit, not the 40S, is responsible for the resistance to the antibiotic (Rao and Grollman, 1967; Jiménez, Littlewood and Davies, 1972). Cycloheximide inhibits not only breakdown, but also reassembly of the polysomes (Colombo, Felicetti and Baglioni, 1966; Wettstein, Noll and Penman, 1964; Stanners, 1966; Marcus and Feeley, 1966), and this suggests that it might inhibit more than one reaction. Indeed the antibiotic has been observed to prevent polysome formation, suggesting an inhibition of initiation (Godchaux, Adamson and Herbert, 1967), and the increase in the puromycin reaction caused by EF-2 treatment suggests an inhibition of translocation (Munro, Baliga and Pronczuck, 1968; Baliga, Pronczuck and Munro, 1969). Indeed, although cycloheximide is a very poor inhibitor of Ac-Phe-tRNA translocation in model systems (Carrasco, Fernández-Puentes and Vázquez, 1976), it very effectively inhibits EF-2-dependent translocation of peptidyl-tRNA in cell-free systems (Baliga, Cohen and Munro, 1970; Obrig et al., 1971; Barbacid, Fresno and Vázquez, 1975). Furthermore, a model system has been developed in which there is a K^+-induced translocation of peptidyl-tRNA by eukaryotic polysomes in the absence of EF-2; cycloheximide is also a strong inhibitor of this non-enzymic translocation by ribosomes from a yeast sensitive strain (Barbacid, Fresno and Vázquez, 1975; Carrasco, Jiménez and Vázquez, 1976) but does not affect a similar reaction by ribosomes from a cycloheximide-resistant mutant (Jiménez, Carrasco and Vázquez, 1977). One protein from the larger ribosomal subunit has been found to be altered in ribosomes from a cycloheximide-resistant fungus (_Podospora anserina_) (Begueret, Perrot and Crouzet, 1977) and yeast (_Schizosaccharomyces pombe_) (Coddington and Fluri, 1977).

Cycloheximide has even been postulated as an inhibitor of chain termination (Godchaux, Adamson and Herbert, 1967; Rajalakshmi et al., 1971). Indeed, by measuring the

Table 12. Comparative effects of some inhibitors of translation that are active on eukaryotic systems

Compound	Inhibition on some systems of initiation	Action on the ribosome	Action on EF-2	Inhibition of enzymic translocation
Diphtheria toxin	-	-	+	+
PA toxin	-	-	+	+
Cycloheximide	+	+	-	+
Pederine	+	+	-	+
Cryptopleurine	-	+	-	+
Tylophorine	-	+	-	+
Tylocrebrine	-	+	-	+
Emetine	-	+	-	+
Tubulosine	-	+	-	+

incorporation of methionine into the N-terminal and internal positions of nascent peptide chains,and by determining the size distribution of polysomes in the presence of cycloheximide, it has been proposed that the drug inhibits initiation, elongation, and termination and that initiation is the most sensitive phase (Oleinick, 1977).

Besides cycloheximide there is an increasing number of compounds which prevent translocation by eukaryotic systems by interacting with the ribosome. Their effects are briefly summarized in Table 12.

Table 12 (continued)

Inhibition of non-enzymic translocation	Cross-resistance with cycloheximide	Cross-resistance with cryptopleurine	Cross-resistance with emetine
-	-	-	-
-	-	-	-
+	+	-	-
+	-	-	-
-	-	+	+
-	-	+	+
-	-	+	+
-	-	+	+
-	-	+	+

4.4 Pederine

Pederine is a toxic subtance accumulated in the coleoptera Paederus fuscipes whose structure has already been established (Cardani et al., 1965) (Fig. 51). It inhibits protein synthesis in eukaryotic cells (Brega et al., 1968; Perani et al., 1968; Tiboni, Parisi and Ciferri, 1968; Jacobs-Lorena, Brega and Baglioni, 1971) and cell-free systems (Tiboni, Parisi and Ciferri, 1968; Jacobs-Lorena, Brega and Baglioni, 1971). Pederine acts on the ribosome (Tiboni, Parisi and Ciferri, 1968; Jacobs-Lorena, Brega and Baglioni, 1971), causing polysome breakdown at low concentrations, whereas at higher concentrations it prevents

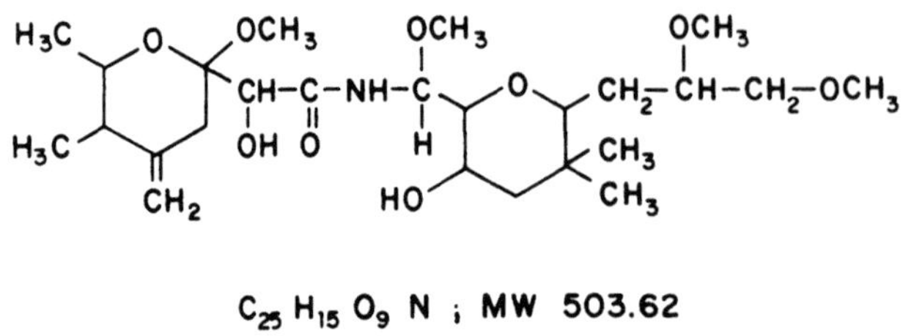

Fig. 51. Chemical structure of pederine

polysome breakdown and only partially inhibits the release of growing peptides by puromycin (Jacobs-Lorena, Brega and Baglioni, 1971). It was therefore suggested that pederine might act at more than one site on the ribosome by inhibiting mainly initiation and to a lesser extent translocation (Jacobs-Lorena, Brega and Baglioni, 1971). Studies with model systems (Carrasco, Fernández-Puentes and Vázquez, 1976) and with yeast polysomes (Barbacid, Fresno and Vázquez, 1975) have shown that pederine blocks translocation. The effect of pederine appears to be located on the 60S ribosome subunit, since it inhibits the K^+-induced non-enzymic translocation of the 3' terminal end of peptidyl-tRNA which takes place in the absence of EF-2 and GTP (Jiménez, Carrasco and Vázquez, 1977). Although cycloheximide and pederine have a similar overall effect, they act at different sites on the 60S ribosomal subunit since pederine also blocks the K^+-induced non-enzymic translocation by ribosomes from a cycloheximide-resistant mutant (Jiménez, Carrasco and Vázquez, 1977) (Table 12).

4.5 Cryptopleurine and the Tylophora Alkaloids (Tylocrebrine and Tylophorine)

Cryptopleurine and the Tylophora alkaloids (tylocrebrine and tylophorine) are closely related in their chemical structure

(Fig. 52), inhibitory spectra, and mode of action. They are active in blocking protein synthesis in eukaryotic cells, tumors (Donaldson, Atkinson and Murray, 1968), and cell-free systems (Haslam, Davey and Linnane, 1968). The alkaloids block the EF-2-dependent translocation step in model systems with ribosomes (Huang and Grollman, 1972b; Carrasco, Fernández-Puentes and Vázquez, 1976) and in polysomes (Jiménez, Carrasco and Vázquez, 1977). Cryptopleurine-resistant mutants of *Saccharomyces cerevisiae* have been isolated (Skogerson, McLaughlin and Wakatama, 1973; Grant, Sánchez and Jiménez, 1974). The resistance is ribosomal (Skogerson, McLaughlin and Wakatama, 1973) and caused by an altered 40S ribosomal subunit (Grant, Sánchez and Jiménez, 1974). Mutations to cryptopleurine resistance are allelic and dominance of sensitivity over resistance may be observed in vivo in heterozygous diploid strains; high resistance to emetine and tubulosine is observed in some cryptopleurine-resistant mutants but not in others (Sánchez, Vázquez and Jiménez, 1977). Resistance to tubulosine and emetine was also observed in mutants of Chinese hamster ovary cells resistant to cryptopleurine and tylocrebrine (Gupta and Siminovitch, 1977a).

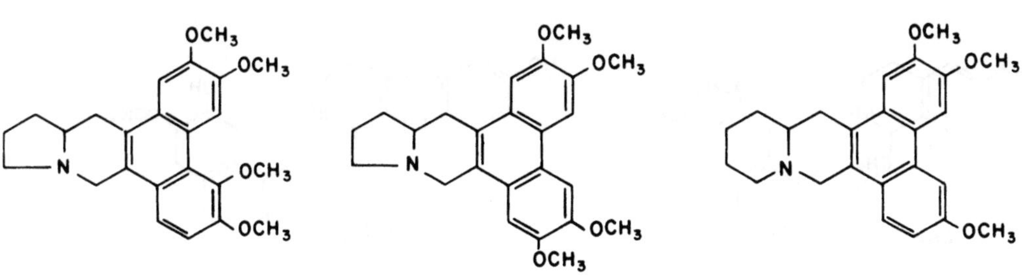

$C_{24}H_{27}O_4N$; MW 393.47 — **Tylocrebrine**

$C_{24}H_{27}O_4N$; MW 393.47 — **Tylophorine**

$C_{24}H_{27}O_3N$; MW 377.47 — **Cryptopleurine**

Fig. 52. Cryptopleurine and the Tylophora alkaloids

Thus, the K^+-induced non-enzymic translocation, in which the 60S subunit is mainly involved, is not affected by the Tylophora alkaloids (Jiménez, Carrasco and Vázquez, 1977) (Table 12). However, at higher concentrations the Tylophora alkaloids also interact with the 60S ribosomal subunit (Battaner and Vázquez, 1971) and inhibit peptide bond formation (Barbacid, Fresno and Vázquez, 1975).

4.6 Emetine and Tubulosine

The alkaloid emetine is the main active product obtained from the roots of *Cephaelis ipecacuanha*; extracts of this root have been in therapeutic use for centuries, since emetine is active on eukaryotic cells and is mainly useful in the treatment of amebiasis. The chemical structure of emetine initially proposed (Fig. 53) has been confirmed by synthesis (Buttersby and Turner, 1960).

The alkaloid tubulosine is obtained from bark extracts of *Pogonopus tubulosus* and *Alangium lamarckii*. Its chemical structure has been elucidated and shown to be closely related to that of emetine (Brauchli et al., 1964; Popelak, Haack and Spingler, 1966) (Fig. 53).

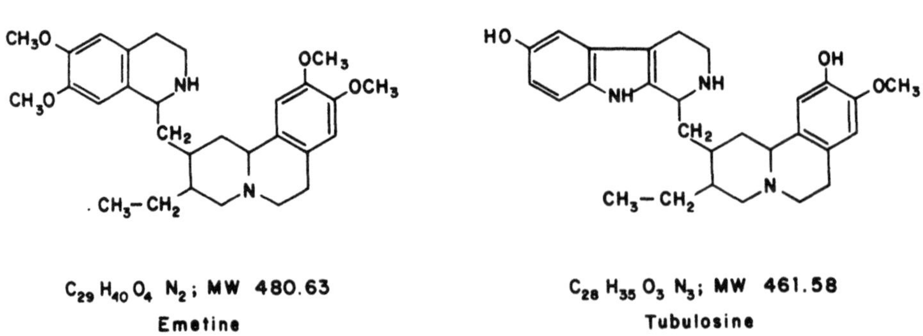

Fig. 53. Chemical structures of emetine and tubulosine

Both emetine and tubulosine selectively prevent protein synthesis in eukaryotic cells and cell-free systems (Grollman, 1966, 1967b, 1968; Farkas et al., 1974; Carrasco, Jiménez and Vázquez, 1976). These alkaloids are very effective inhibitors of EF-2-dependent translocation by polysomes (Barbacid, Fresno and Vázquez, 1975; Carrasco, Jiménez and Vázquez, 1976; Jiménez, Carrasco and Vázquez, 1977). The only significant difference that has been described concerning the mode of action of these alkaloids is that tubulosine has been observed to block EF-2-dependent translocation of Ac-Phe-tRNA by ribosomes (Carrasco, Jiménez and Vázquez, 1976) whereas emetine is not an active inhibitor in this system (Carrasco, Fernández-Puentes and Vázquez, 1976).

Emetine and tubulosine, unlike cycloheximide and pederine, do not affect the K^+-induced non-enzymic translocation by eukaryotic ribosomes (Carrasco, Jiménez and Vázquez, 1976; Jiménez, Carrasco and Vázquez, 1977). Furthermore, cycloheximide-resistant polysomes are sensitive to emetine and tubulosine, clearly suggesting a different site of action for the antibiotic and the two alkaloids (Jiménez, Carrasco and Vázquez, 1977). On the other hand polysomes resistant to cryptopleurine are also resistant to emetine, tubulosine, tylophorine, and tylocrebrine since the five drugs interact with the same site on the 40S ribosomal subunit (Sánchez, Vázquez and Jiménez, 1977). This was confirmed with ribosomes of Chinese hamster ovary cells resistant to cryptopleurine since the resistance is expressed in the 40S ribosomal subunits and there is cross-resistance to emetine to tubulosine (Gupta and Siminovitch, 1977b). All the above results suggest that emetine and tubulosine might act on the 40S ribosomal subunit on the same site as cryptopleurine, tylophorine, and tylocrebrine (Table 12).

4.7 Viomycin and Capreomycin

Viomycin (synonyms: celiomycin, florimycin, tuberactinomycin B, and viocin) is a basic antibiotic which is produced by Streptomyces puniceus (Finlay et al., 1951) and belongs to a group of closely related antibiotics known as tuberactinomycins. The chemical structures of the different tuberactinomycins including viomycin have already been elucidated (Nomoto et al., 1977) (Fig. 54).

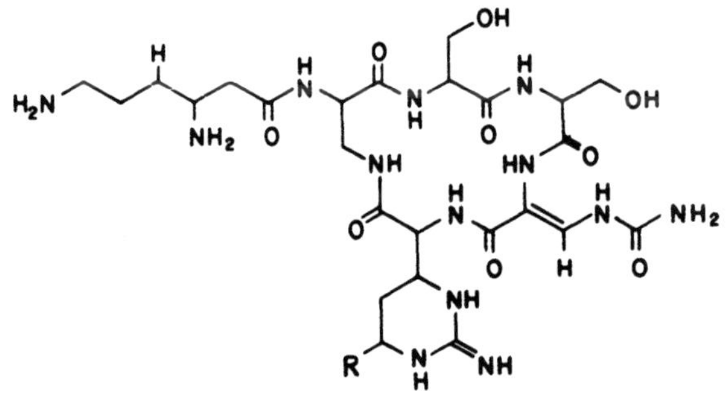

Viomycin: R = OH ($C_{25}H_{43}O_{10}N_{13}$; MW 685.70)

Tuberactinomycin O: R = H ($C_{25}H_{43}O_{9}N_{13}$; MW 669.70)

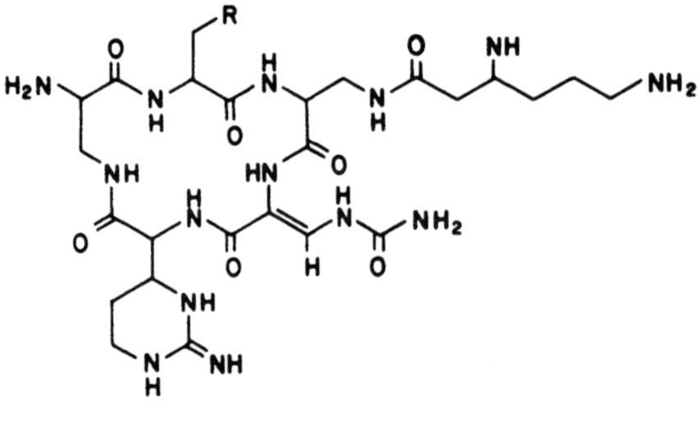

Capreomycin 1A: R = -OH ($C_{25}H_{44}N_{14}O_{8}$; MW 668.73)

Capreomycin 1B: R = -H ($C_{25}H_{44}N_{14}O_{7}$; MW 652.73)

Fig. 54. Chemical structures of viomycin and the capreomycins

Capreomycin (synonym capromycin) is a complex antibiotic produced by Streptomyces capreolus (Herr, 1962), having four components (capreomycin IA, IB, IIA and IIB (Fig. 54); they are related to viomycin and their chemical structures have been elucidated recently (Nomoto et al., 1977).

Viomycin and the capreomycins are mainly active on mycobacteria and to a lesser extent on gram-positive bacteria; for this reason viomycin and the capreomycin complex have a certain use in medicine as an antituberculous agent. Little is known about the mode of action of capreomycin IA, IB and II but they are assumed to have a mode of action similar to that of viomycin.

Viomycin inhibits protein synthesis in intact bacteria (Jordan, Yamamura and McKague, 1969) and bacterial cell-free systems with endogenous mRNA (Tanaka and Igusa, 1968) or synthetic polynucleotide (Alexander, Jordan and McKague, 1969). Viomycin inhibits AUG- and phage f2 RNA-dependent binding of f-Met-$tRNA_F$ to 30S ribosome subunits and was therefore postulated as an inhibitor of initiation (Liou and Tanaka, 1976). Viomycin is also an effective inhibitor of elongation and specifically prevents EF-G-dependent translocation of Ac-Phe-tRNA (Liou and Tanaka, 1976). This dual effect of viomycin might be due to two different types of interaction of the antibiotic with the ribosome. Hence there are two types of multiple step selected viomycin-resistant mutants of Mycobacterium smegmatis, one with altered 30S ribosomal subunits and the other with altered 50S subunits (Yamada et al., 1972). Ribosomes from the viomycin-resistant mutants with altered 30S ribosomal subunits are also resistant to kanamycin and streptomycin (Yamada et al., 1974), although viomycin resistance is mapped at a locus different from streptomycin and kanamycin loci (Yamada et al., 1976). On the other hand ribosomes from the viomycin-resistant mutants with altered 50S subunits are sensitive to streptomycin and kanamycin (Yamada et al., 1974). These results are in accordance with the possibility that the viomycin interaction with the 30S subunit might be responsible for the inhibitory effect of the antibiotic in initiation,whereas an interaction of the antibiotic with the 50S subunit might cause the inhibition of translocation. It was indeed shown recently that (^{14}C) tuberactinomycin O, that is closely related to viomycin, binds to both the 30S and the 50S ribosomal subunits; the affinity of the antibiotic for the 50S subunit is higher (Misumi, Tanaka and Shiba, 1978). Binding of (^{14}C) tuberactinomycin O to ribosomes and 50S and 30S subunits is equally inhibited by viomycin (Misumi, Tanaka and Shiba, 1978).

The inhibitory effect of viomycin on translocation has been confirmed in purified E. coli polysomes engaged in polypeptide synthesis (Modolell and Vázquez, 1977). The antibiotic does not impair the EF-G plus ribosome-dependent GTP hydrolysis or the formation of the guanosine-nucleotide EF-G·ribosome·fusidic acid complex or the non-enzymic binding of N-acetyl-Phe-tRNA to the ribosomal donor site, but strongly blocks non-enzymic translocation and the release of N-acetyl-Phe-tRNA from the ribosomal acceptor site promoted by depletion of NH_4^+ ions (Modolell and Vázquez, 1977). At rather high concentrations viomycin partially inhibits the EF-G plus ribosome-dependent GTP hydrolysis and the turnover of the GDP-EF-G-ribosome-fusidic acid complex, but these effects are probably unrelated to the inhibition of ribosomal translocation (Modolell and Vázquez, 1977). This is not surprising since there is evidence that viomycin interaction with the ribosome modifies the iodination of at least five proteins of the 30S ribosomal subunit and four proteins of the 50S ribosomal subunit (Martínez, Vázquez and Modolell, 1978).

4.8 Hygromycin B

Hygromycin B is an aminoglycoside antibiotic, obtained from cultures of Streptomyces hygroscopicus, whose chemical structure has been elucidated (Neuss et al., 1970). Hygromycin B has some unusual properties clearly differing from other aminoglycoside antibiotics in its chemical structure (Fig. 55) and inhibitory spectra since it is active on prokaryotic and eukaryotic cells and cell-free systems.

Hygromycin B induces misreading as assayed in bacterial (Davies, Gorini and Davis, 1965) and eukaryotic (González et al., 1978) cell-free systems for polypeptide synthesis directed by synthetic polynucleotides (Davies, Gorini and Davis, 1965). Furthermore hygromycin B inhibits the EF-G-dependent translocation of Ac-Phe-tRNA or peptidyl-tRNA and impairs the exit of Ac-Phe-tRNA from the ribosomal A-site owing to depletion of NH_4^+ ions. The inhibitory effect of hygromycin B in translocation appears to be a consequence of its interaction with the ribosome, since the antibiotic prevents non-enzymic synthesis of polyphenylalanine

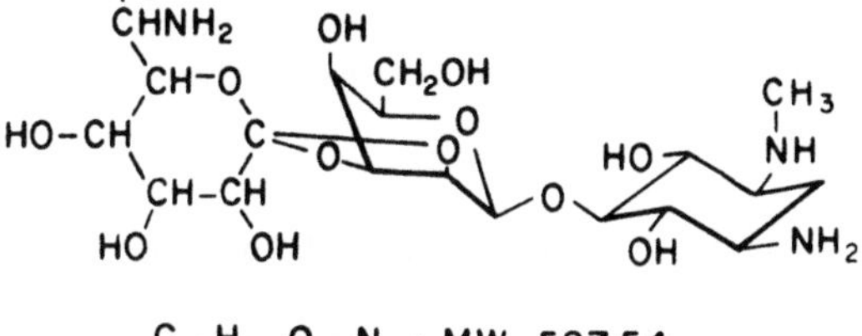

$C_{20}H_{37}O_{13}N_3$; MW 527.54

Fig. 55. Chemical structure of hygromycin B

(Cabañas, Vázquez and Modolell, 1978). Moreover, in a crude poly(U)-directed system inhibition of translocation and of polypeptide synthesis and induction of misreading occur simultaneously (Cabañas, Vázquez and Modolell, 1978).

Hygromycin B is also a very active inhibitor of polypeptide translocation in eukaryotic systems, acting by blocking elongation. Similarly as with some other inhibitors of translocation that act on the 40S ribosomal subunit (emetine, tubulosine, and cryptopleurine) the inhibitory extent of translocation by hygromycin B is decreased by increasing concentrations of elongation factor EF-2. However, hygromycin B acts at a different site on the 40S ribosomal subunit since it prevents translocation by ribosomes resistant to cryptopleurine. Hygromycin B also induces misreading in eukaryotic systems (González et al., 1978).

4.9 Streptomycin and Other Aminoglycoside Antibiotics

Recent studies (Cabañas, Vázquez and Modolell, 1978) have shown that, besides hygromycin B, other misreading-inducing aminoglycoside antibiotics can interfere with ribosomal translocation. In a model system, the EF-G plus GTP dependent reaction of A-site bound Ac-Phe-tRNA with puromycin is inhibited over 80% by 3 µM neomycin B, 10 µM neomycin C, and 30 µM paromomycin, ribostamycin, kanamycin A, or kanamycin B. Tobramycin, gentamicin C, gentamicin C_{1A}, sisomicin and verdamicin inhibit over 50% at 5 µM but inhibitions higher that 70% are not observed even at 0.5 mM. Neamine requires approximately 20 µM for 50% inhibition and 0.5 mM for 80% inhibition, and streptomycin and bluensomycin have smaller effects, causing 20 to 40% inhibition even at high concentrations (0.1 mM). A relatively weak inhibition of translocation by 0.5 mM streptomycin had been previously reported by Igarashi, Ishitsuka and Kaji (1969). Moreover, unpublished work in our laboratory with an in vitro system containing endogenous E. coli polysomes and crude S100 extract has shown that 0.1 mM neomycin B added to this sytem actively synthesizing polypeptide inhibits the synthesis and retains peptidyl-tRNA in a puromycin unreactive position. Thus it seems likely that, at least for the most potent aminoglycoside antibiotic inducers of misreading, the inhibition of translocation plays an important role in the mechanism of inhibition of protein synthesis by these antibiotics.

5. Miscellaneous Inhibitors of Elongation

5.1 Macrolide Antibiotics of the Erythromycin, Methymycin, and Lancamycin Groups

Within the family of the macrolide antibiotics we can distinguish the carbomycin and spiramycin groups which are inhibitors of peptide bond formation, and the antibiotics of the erythromycin, methymycin, and lancamycin groups which have a different mode of action and will be discussed here. The best-known antibiotics of these groups are those indicated in Table 13 and Figure 34. The chemical structures of their sugar moieties are shown in Figure 35.

The mode of action of the macrolide antibiotics of the erythromycin, methymycin, and lancamycin groups has been reviewed recently (Vázquez, 1975a; Oleinick, 1975). Erythromycin inhibits protein synthesis in intact bacteria and bacterial cell-free

Table 13. Macrolide antibiotics of the erythromycin, methymycin, and lancamycin groups

(a) Source of antibiotic complexes and the individual components

Antibiotic complex (synonym)	Individual components (synonym)	Producer species of streptomyces
	Chalcomycin	S. bikinensis
	(Bandamycin B)	S. albogriseolus
	(Miconomycin)	
Erythromycin	Erytromycin A	S. erythreus
(Erytrocin)	Erytromycin B	
(Ilotycin)	Erytromycin C	
Kujimycin	Kujimycin A	S. spinichromogenes
	Kujimycin B	var. kujimiceticus
	(Lancamycin)	S. violaceoniger
	(Lancamycin)	
	Methymycin	S. eurocidicus
	Narbomycin	S. narbonensis
	Neomethymycin	S. venezuelae
	Oleandomycin	S. antibioticus
	(Amimycin)	S. olivochromogenes
	(Matromycin)	
	(Romicil)	
	(Landomycin)	
	Picromycin	S. felleus
		S. venezuelae

Data taken from Vāzquez (1975a).

Table 13 (continued)

(b) Classification of the sugar moieties linked to the aglycone ring

Group	Characteristics	Members of the group	Sugars linked to the aglycone	
Erythromycin	a	Erythromycins	-desosamine	-cladinose or -mycarose
		Neospiramycins	-mycaminose	-isomycamine
		Oleandomycin	-desosamine	-oleandrose
		Megalomycin A	-desosamine	-mycarose
Methymycin	b	Methymycin	-desosamine	
		Forocidins	-mycaminose	
		Narbomycin	-desosamine	
		Neomethymycin	-desosamine	
		Picromycin	-desosamine	
		Demycarosylturimycin	-mycaminose	
Lancamycin	c	Chalcomycin	-mycinose	-chalcose
		Neutramycin	-mycinose	-chalcose
		Lancamycin (synonym Kujimycin B)	-acetylarcanose	-chalcose (synonym lancavose)
		Kujimycin A	-arcanose	-chalcose

[a] Containing the moieties -amino sugar and either -amino sugar or -non nitrogenous sugar linked to the aglycone.

[b] Containing the moiety -amino sugar linked to the aglycone.

[c] Containing the moieties -non nitrogenous sugar and -non nitrogenous sugar linked to the aglycone.

Data taken from Vâzquez (1975a).

systems (Vázquez, 1966a; Taubman, Young and Corcoran, 1963; Wolfe and Hahn, 1964). One molecule of erythromycin binds to sensitive ribosomes on the larger subunit (Mao and Putterman, 1969; Taubman, Young and Corcoran, 1963; Oleinick and Corcoran, 1969; Teraoka, 1970; Fernández-Muñoz and Vázquez, 1973a,b). Erythromycin prevents chloramphenicol (Vázquez, 1966b), lincomycin (Chang and Weisblum, 1967), and spiramycin III binding (Vázquez, 1967b) and vice versa (Wilhelm, Oleinick and Corcoran, 1968). Furthermore erythromycin prevents the inhibitory effects of chloramphenicol and lincomycin on the peptidyl transferase (Celma, Monro and Vázquez, 1970; Chang and Weisblum, 1967; Fernández-Muñoz and Vázquez, 1973c). These results and the reported inhibition by erythromycin of the puromycin (Tanaka, Teraoka and Tamaki, 1971; Černá, Jonák and Rychlik, 1971; Černá, Rychlik and Pulkrabek, 1969; Rychlik and Černá, 1972) and the fragment reaction (Mao and Robishaw, 1972) assays when certain donor substrates were used, suggested that erythromycin might be an inhibitor of peptide bond formation. However, erythromycin does not inhibit peptide bond formation in most model systems. Some studies in intact cells (Cundliffe and McQuillen, 1967) and cell-free systems (Cannon and Burns, 1971; Igarashi, Ishitsuka and Kaji, 1969; Oleinick and Corcoran, 1970; Tanaka, Otaka and Kaji, 1973) indicated that erythromycin might be an inhibitor of translocation; however, erythromycin does not inhibit translocation in most model systems. Binding experiments have shown that erythromycin cannot interact with polysomes (Pestka, 1974; Tai, Wallace and Davies, 1974) since the nascent peptide moiety of the peptidyl-tRNA bound to either the P- or the A-site inhibits the binding of the antibiotic to ribosomes (Contreras and Vázquez, 1977a). The antibiotic does not therefore inhibit polypeptide synthesis by polysomes (Pestka, 1972b) but interacts with 50S subunits and free ribosomes and initiation takes place; however, the antibiotic blocks elongation prior to polysome formation, most probably by steric hindrance to the growing peptidyl chain (Contreras and Vázquez, 1977a). Thus erythromycin blocks peptide bond formation in the puromycin assay when oligopeptidyl-tRNA of a certain length is the donor substrate (e.g., pentalysyl- or Ac-Pro-Gly- or Ac-Phe-Pro-tRNA) (see review by Vázquez, 1975a) but it has a very small inhibitory effect or even enhances peptide bond formation when the donor substrate is lysyl- or dilysyl- or trilysyl- or Ac-Gly-Gly- or Ac-Phe- or f-Met-tRNA (see review by Vázquez, 1975a). These data suggest that the minimum length of the nascent oligopeptide chain whose growth is blocked by erythromycin with natural mRNA might be from di- to pentapeptide depending on the amino acid composition. This proposal is in agreement with and defines more precisely the previous hypothesis that erythromycin stops protein synthesis soon after initiation (Tai, Wallace and Davis, 1974). Thus ribosomes with erythromycin bound could carry out the first rounds of elongation. When the nascent peptide reaches a certain length (two to five residues, depending on its amino acid composition) the antibiotic bound prevents any further elongation of the oligopeptide chain by steric hindrance. The addition of the sequential amino acids takes place in peptide bond formation on the A-site and does not appear to be inhibited by erythromycin in intact bacteria (Cundliffe and McQuillen, 1967). However, the antibiotic bound to the ribosome interferes with the new oligopeptidyl-tRNA formed while it is moving from the A-site to the P-site and is therefore blocked between both sites or released. Thus translocation is not completed and an empty A-site and P-site occupied by uncharged tRNA might be the result (Tanaka, Otaka and Kaji, 1973).

Macrolide antibiotics are mutually exclusive in their binding to ribosomes (Vâzquez, 1967b; Fernândez-Muñoz et al., 1971; Wilhelm, Oleinick and Corcoran, 1968; Mao and Putterman, 1969; Pestka, Nakagawa and Omura, 1974) and there is most probably a common site on the ribosome involved in the interaction of the aglycone ring of the different macrolides. Thus ribosomes from the erythromycin-producing strain of Streptomyces erythreus have very low affinity for all the macrolide antibiotics tested (Teraoka and Tanaka, 1974). Whereas macrolides of the spiramycin and carbomycin groups block peptide bond formation, antibiotics of the erythromycin, methymycin and lancamycin groups appear to act like erythromycin (see above). As in spiramycin resistance (Tanaka et al., 1971), protein L4 is altered in ribosomes from some mutants of E. coli resistant to erythromycin and the affinity of the antibiotic is drastically reduced (Otaka et al., 1970, 1971; Dekio et al., 1970; Pardo and Rosset, 1974). However, protein L22 is altered in another E. coli mutant resistant to erythromycin and yet the antibiotic binds to the ribosome (Wittmann et al., 1973; Pardo and Rosset, 1977). Furthermore a third mutation leading to erythromycin resistance seems to affect mainly initiation, but the ribosomal component altered has not been identified (Pardo and Rosset, 1977). On the other hand, induced and constitutive resistance to the antibiotic in Staphylococcus aureus has been associated with the modification of 23S tRNA in which formation of dimethyladenine has been observed (Lai and Weisblum, 1971; Lai et al., 1973). Similarly there is a specific methylation of the 23S RNA of ribosomes from Streptomyces erythreus (B. Weisblum, personal communication), the producer microorganisms of erythromycin.

Protein L16 appears to be required for erythromycin binding to bacterial ribosomes, since the loss of activity of the particles to bind erythromycin closely parallels their loss of protein L16 (Bernabeu, Vázquez and Ballesta, 1977). Nevertheless erythromycin binding to free protein L15 has been reported, whereas it was confirmed in reconstitution experiments that protein L16 is required for erythromycin binding to the 50S ribosomal subunit and no binding was observed in 50S derived cores deprived of protein L16 even if they contain protein L15 (Teraoka and Nierhaus, 1978).

5.2 Antibiotics of the Streptogramin B Group

Antibiotics of the streptogramin B group have been reviewed recently (Vázquez, 1975b; Tanaka, 1975). They are antibacterial compounds produced by Streptomyces which are mainly active on gram-positive bacilli. In most cases, streptogramin B antibiotics are produced in mixture with streptogramin A antibiotics which are inhibitors of translation acting on peptide bond formation (Table 8). The chemical structure of streptogramin B antibiotics has already been established (Fig. 56) (Vázquez, 1975b; review).

Antibiotics of the streptogramin B group (viridogrisein, staphylomycin S, streptogramin B and synonyms) inhibit protein synthesis in intact bacteria and bacterial cell-free systems (Vázquez, 1966a, 1966c; Ennis, 1965a, 1965b; Yamaguchi and

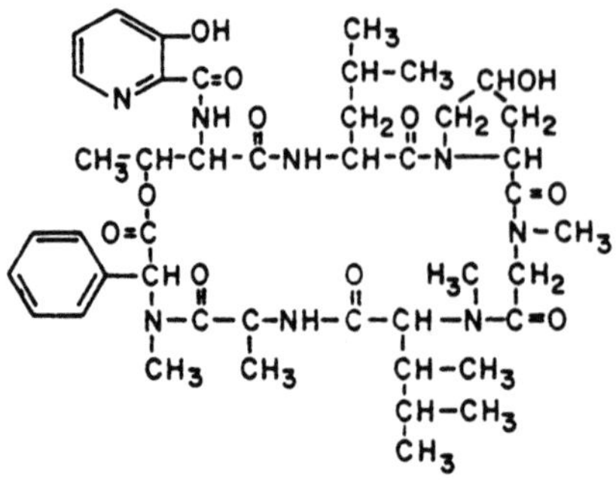

$C_{44}H_{62}O_{11}N_8$; MW 879.04
Viridogrisein (synonym etamycin)

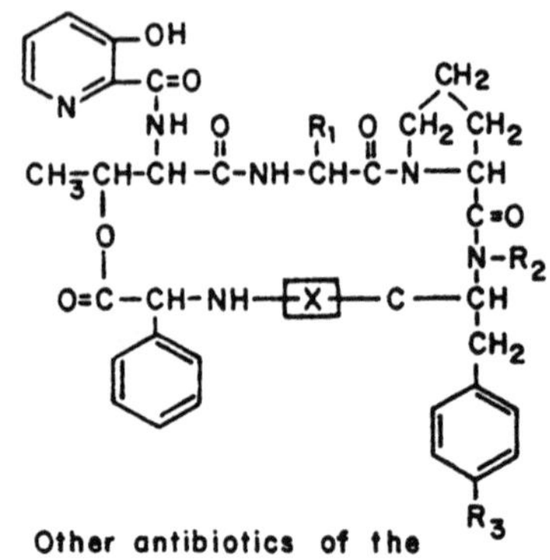

Other antibiotics of the streptogramin B group

Antibiotic (synonym)	R_1	R_2	R_3	-[X]-
Streptogramin B (Ostreogrycin B) (Mikamycin B) (Synergistin B·1) (Vernamycin Bα) (Pristinamycin I_A)	$-CH_2-CH_3$	$-CH_3$	$-N(CH_3)_2$	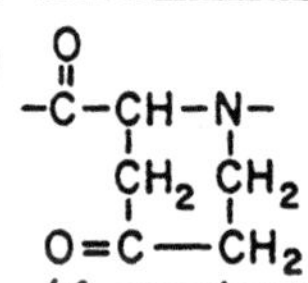 (4-oxopipecolic acid)
Ostreogrycin B_1 (Pristinamycin I_C) (Vernamycin B_γ)	$-CH_3$	$-CH_3$	$-N(CH_3)_2$	4-oxopipecolic acid
Ostreogrycin B_2 (Pristinamycin I_B) (Vernamycin B_β)	$-CH_2-CH_3$	$-CH_3$	$-NH-CH_3$	4-oxopipecolic acid
Ostreogrycin B_3	$-CH_2-CH_3$	$-CH_3$	$-N-(CH_3)_2$	3-hydroxy-4-oxopipecolic acid
Patricin A	$-CH_2-CH_3$	$-CH_3$	$-H$	proline
Patricin B	$-CH_2-CH_3$	$-CH_3$	$-H$	pipecolic acid
Vernamycin B_δ	$-CH_3$	$-CH_3$	$-NH-CH_3$	4-oxopipecolic acid
Vernamycin C	$-CH_2-CH_3$	$-CH_3$	$-N(CH_3)_2$	aspartic acid
Virginiamycin S	$-CH_2-CH_3$	$-CH_3$	$-H$	4-oxopipecolic acid
Virginiamycin S_2	$-CH_2-CH_3$	$-H$	$-H$	4-hydroxypipecolic
Virginiamycin S_3	$-CH_2-CH_3$	$-CH_3$	$-H$	3-hydroxy-4-oxopipecolic acid
Virginiamycin S_4	$-CH_3$	$-CH_3$	$-H$	4-oxopipecolic acid

Streptogramin B: $C_{45}H_{54}O_{10}N_8$; MW 867
Virginiamycin S: $C_{43}H_{49}O_{10}N_7$; MW 823.91

Fig. 56. Antibiotics of the streptogramin B group

Tanaka, 1964; Laskin and Chan, 1965). These antibiotics interact with the 50S ribosomal subunit (Vázquez, 1966b, 1967b; Ennis, 1966, 1974; Chang, Siddhikol and Weisblum, 1969) but do not bind to polysomes (Ennis, 1974) since peptidyl-tRNA bound to either the P- or the A-site prevents the attachment of these antibiotics (Contreras and Vázquez, 1977a). Streptogramin B antibiotics inhibit erythromycin and spiramycin III binding to the ribosome (Vázquez, 1967b; Wilhelm, Oleinick and Corcoran, 1968). Conversely streptogramin B binding to ribosomes is inhibited by all the different macrolide antibiotics that were tested (Ennis, 1974). Like erythromycin, streptogramin B antibiotics do not inhibit the puromycin reaction with the donor substrate f-Met-tRNA but partially inhibit the reaction with the donor substrate f-Met-Ala-tRNA (Kubota, Okuyama and Tanaka, 1972). Also as in the case of erythromycin, protein L16 is required for the interaction of streptogramin B antibiotics with the 50S ribosomal subunit (Béthune and Nierhaus, 1978). All the evidence suggests that like erythromycin the streptogramin B antibiotics block the growth of the nascent polypeptide chain by steric hindrance (Contreras and Vázquez, 1977a). There is a remarkable synergism between streptogramin B and streptogramin A antibiotics in gram-positive bacteria. This appears to be due to the enhancement of streptogramin B binding to the ribosome by streptogramin A (Contreras and Vázquez, 1977b), the streptogramin B-ribosome complex being more stable in the presence of streptogramin A (Ennis, 1974).

5.3 Spectinomycin (Synonyms Actinospectacin and Trobicin)

Spectinomycin was isolated initially from cultures of *Streptomyces spectabilis* (Mason, Dietz and Smith, 1961) and subsequently from *S. hygroscopicus* and *S. flavopersicus*. It is an antibacterial antibiotic which is active mainly on gram-positive bacteria and has certain medicinal and veterinary uses. Chemical studies showed that spectinomycin is an aminoglycoside antibiotic but lacks the deoxystreptamine or streptamine moiety characteristic of some other aminoglycoside antibiotics (Fig. 57).

Spectinomycin inhibits protein synthesis but, unlike those aminoglycoside antibiotics having the streptamine or deoxystreptamine moiety, does not cause misreading (Davies, Anderson and Davis, 1965, 1967). Resistance to spectinomycin resides on the 30S ribosomal subunit (Davies, Anderson and Davis, 1965; Bollen et al., 1969) and is due to an alteration of protein S5 (Wittmann et al., 1973; Bollen *et al.*, 1969; Funatsu, Nierhaus and Wittmann-Liebold, 1972). Since protein S5 is somehow involved in EF-Tu and EF-G-dependent GTP hydrolysis, spectinomycin has been proposed as an inhibitor of translocation (Marsh and Parmeggiani, 1973). Further work with bacterial

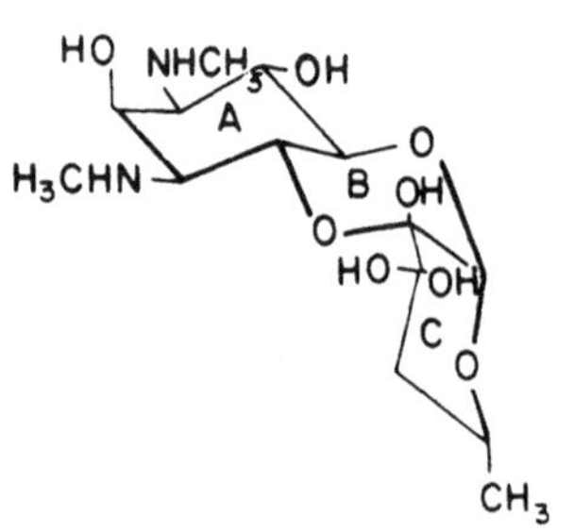

$C_{14}H_{24}O_7N_2$; MW 332.35

Fig. 57. Chemical structure of spectinomycin

protoplast has shown that in vivo spectinomycin can prevent polysome breakdown and inhibit translocation (Burns and Cundliffe, 1973). However in in vitro translating systems spectinomycin fails to inhibit preformed polysomes except at rather high concentrations of the antibiotic (Wallace, Tai and Davis, 1974). It has therefore been postulated that spectinomycin might only inhibit the early rounds of translocation prior to polysome formation (Wallace, Tai and Davis, 1974).

The cold-sensitivity of spc^r mutants can be relieved by a secondary alteration in either protein S2, S3, or S5 (Nashimoto and Uchida, 1975). On the other hand an E. coli mutant has been isolated which shows spectinomycin dependence; it is a double mutant with one mutation in the str A region, affecting protein S5 and conferring a spectinomycin resistant phenotype on the strain, and the other in the rif region, conferring spectinomycin dependence. There are a number of revertants for spectinomycin dependence showing alteration in either protein S4 or L5 (Dabbs, 1977).

5.4 Bottromycin A_2

Bottromycin A_2 is the best known component of the bottromycin complex of polypeptide antibiotics produced by Streptomyces bottropensis, S. canadensis and Kitasatoa purpurea (Waisvisz et al., 1957). It is an antibacterial antibiotic which is mainly active on Mycoplasma species and also on gram-positive bacteria and mycobacteria.

The chemical structure of bottromycin A_2 has recently been revised (Takahashi et al., 1976) (Fig. 58). The antibiotic

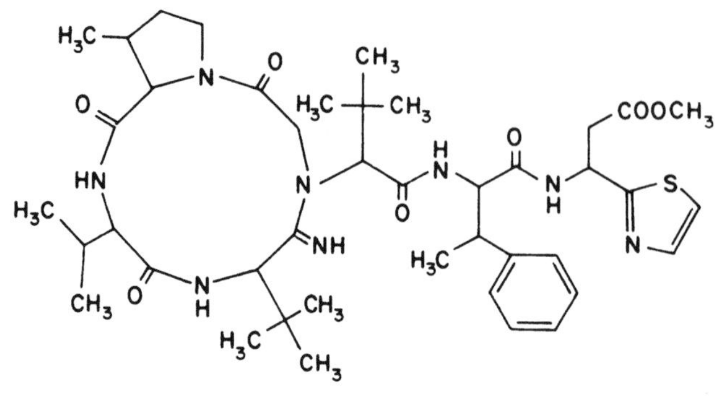

Fig. 58. Chemical structure of bottromycin A_2

inhibits protein synthesis in bacteria and bacterial cell-free systems (Lin, Tanaka and Umezawa, 1967; Lin and Tanaka, 1968). Bottromycin A_2 acts on the 50S subunit but has no effect on (^{14}C)-chloramphenciol binding (Lin and Tanaka, 1968). Bottromycin A_2 has been reported to have a certain inhibitory effect on the puromycin (Pestka and Brot, 1971) and the fragment reaction (Tanaka, Lin and Okuyama, 1971) assays, but peptide bond formation does not appear to be the main target of the antibiotic (Pestka and Brot, 1971). Indeed bottromycin A_2 is a good inhibitor of the translocation of polylysyl-tRNA attached to polysomes (Lin, Kinoshita and Tanaka, 1968) and f-Met-tRNA in a resolved assay (Tanaka, Lin and Okuyama, 1971). This effect of the antibiotic might be due to its action in releasing the substrate bound to the A-site of the ribosome (Otaka and Kaji, 1976), although there are contradictory reports concerning the effect of the antibiotic on this site. Thus bottromycin A_2 has been reported both to inhibit (Otaka and Kaji, 1976) and to have no effect on EF-Tu-dependent binding of Phe-tRNA to ribosomes (Pestka and Brot, 1971).

5.5 Colicin E3 and Cloacin DF13

Although the bacteriocins colicin E3 and cloacin DF13 (synonym bacteriocin DF13)(MW = 56,000) differ in their origin and spectra, both proteins appear to have the same mode of action. The molecular weights 60,000 (Sidikaro and Nomura, 1973) and 50,000 (Ohno et al., 1977) have been assigned for colicin E3. Colicin E3 is produced by colicinogenic strains of Escherichia coli and has a bactericidal effect on strains of Escherichia and Shigella (Sidikaro and Nomura, 1973), whereas cloacin DF13 is produced by cloacinogenic strains of Enterobacter cloacae and has a killing effect on bacterial strains of Enterobacter and Klebsiella (Stouthamer and Tieze, 1966). However, in cell-free systems the inhibitory spectra of these proteins are wider. Thus ribosomes from Bacillus stearothermophilus and Azotobacter vinelandii are inactivated by colicin E3 (Sidikaro and Nomura, 1973) and ribosomes from Escherichia coli are inactivated by cloacin DF13, but this bacteriocin does not inactivate ribosomes from Bacillus licheniformis (Graaf, Niekus and Klootwijk, 1973). Colicin E3 inhibits protein synthesis (Konisky and Nomura, 1967) by specifically cleaving a fragment of 49 nucleotides from the 3' end of the 16S ribosomal RNA of intact sensitive bacteria (Senior and Holland, 1971; Bowman et al., 1971) and in ribosome suspensions (Boon, 1971; Bowman, Sidikaro and Nomura, 1971). It has been shown that (^{14}C)cloacin free and complexed with its immunity protein binds with similar affinity to membrane vesicles from sensitive cells, but removal of the immunity protein from the complex appears to reduce the transfer of the cloacin molecules from the cell surface to the cytoplasmic membrane (Oudega et al., 1977). In cell-free extracts the ribosomes of the colicinogenic strains are resistant to colicin E3 (Boon, 1971) since these strains produce an "immunity protein" that interacts very strongly with colicin E3, preventing its action (Bowman, Sidikaro and Nomura, 1971; Boon, 1972). Native colicin E3 (MW = 60,000) has been shown to be a complex of protein A and protein B (Fig. 59) (Jakes and Zinder, 1974a). A different fractionation of colicin E3 yields two fractions, T1 and T2 (Lau and Richards, 1976). A recent report has proposed that protein A (MW = 50,000) is the real active colicin E3. It appears that the bacteriocin is bound to protein B (MW = 10,000) that is the immunity protein (Ohno et al., 1977) in the native colicin E3 usually prepared by most workers. T1 fragment (MW = 35,000) appears to be one part of protein A that is devoid of enzymic activity and might be involved in the attachment of colicin E3 to the membrane and fraction T2 can be further fractionated in fragment T2A (MW = 15,000), that is, the protein A moiety with enzymic activity, and protein B (Ohno et al., 1977) (Fig. 59). Protein A is more active in intact bacteria and in

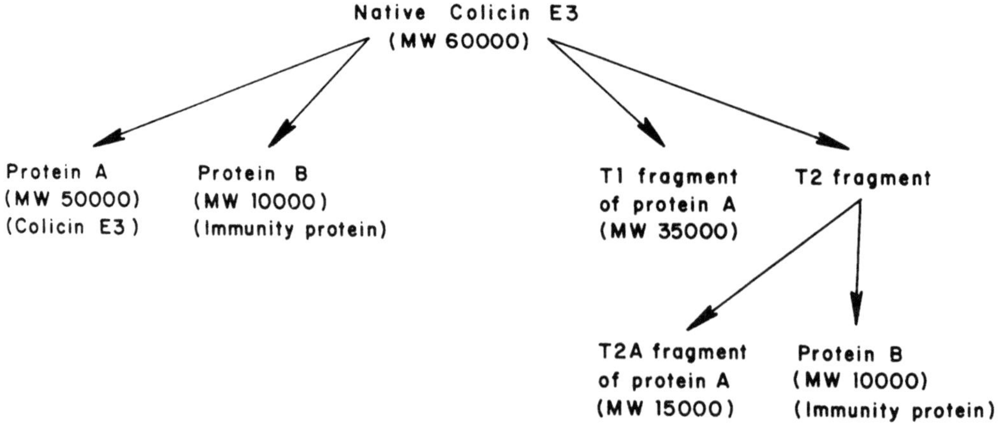

Fig. 59. Structures of native colicin E3, free colicin E3, and immunity protein

cell-free systems than native colicin E3, since it is free from the immunity protein. On the other hand fragment T2A is not bactericidal but is even more active than protein A in cell-free systems (Ohno et al., 1977).

There are conflicting reports concerning substrate specificity for colicin E3: thus 50S and 30S ribosomal subunits have been reported to be required for catalytic inactivation by colicin E3 although the 50S subunit is unaffected (Boon, 1972; Bowman, 1972). The 50S subunits can be replaced in the reaction by 50S-derived cores deprived of proteins L1, L7, L10, L12, L16, L25,and L33 (Sander, 1977b). The antibiotics streptomycin, gentamycin, and tetracycline block this inactivation (Dahlberg et al., 1973). On the other hand it has been reported that protein A (the active component of colicin E3) is active on the 30S ribosomal subunits in the absence of 50S subunits (Ohno, Iwashita and Imahori, 1977).

Since ribosome inactivation by colicin E3 is attained in highly resolved systems using washed ribosomes, it is widely accepted that the bacteriocin inactivates the ribosomes by direct action, although some workers have proposed that this is not so (Ohsumi and Imahori, 1974). The 3' end of 16S RNA has been proposed to be involved in initiation since it is complementary to initiating sequences of different mRNA (Shine and Dalgarno, 1975), binds to protein S1 (Dahlberg and Dahlberg, 1975), which is involved in initiation (Van Dieijen et al., 1975; Szer, Hermoso and Leffler, 1975), and cross-links with initiation factors IF-2 and IF-3 (Traut et al., 1974). Furthermore, the 3' end of 16S RNA cleft by colicin E3 has been reported to be the binding site of proteins S1, S12,and S21 (Nomura, 1976). An inhibitory effect of colicin E3 on initiation might therefore be expected. Indeed colicin E3 has been reported to inhibit binding of f-Met-tRNA$_F$ to the 30S ribosomal subunit (Tai, 1975), although no such inhibition was observed by other workers in a previous report (Turnowsky and Högenauer, 1973). Colicin E3 also interacts with polysomes and inhibits chain elongation (Tai and Davis, 1974; Jakes and Zinder, 1974b). Thus, to a certain extent colicin E3 inhibits non-enzymic (Konisky and Nomura, 1967) and enzymic (Turnowsky and Högenauer, 1973) binding of aminoacyl-tRNA. This inhibitory effect by colicin E3 is only observed at low inhibitory concentrations of Mg^{2++} (5 mM Mg^{2+}) and under these conditions the EF-T plus ribosome dependent GTPase is uncoupled and stimulated 6-7 fold by colicin E3 treatment (Sander, 1977a). Furthermore aminoacyl-tRNA binding to the ribosome protects the latter from interaction with (^{14}C)colicin E3 (Sander, 1977a) and inactivation by colicin E3 (Kaufmann and Zamir, 1973, 1975).

Cloacin DF13, acting catalytically on washed ribosomes or synthesizing systems (Oudega and Graaf, 1976), specifically cleaves 48-49 nucleotides from the 3' terminus of 16S RNA (Graaf, Niekus and Klootwijk, 1973) and can partly inhibit f-Met-tRNA$_F$ (Graaf, Planta and Stouthamer, 1971; Baan et al., 1976) and aminoacyl-tRNA binding (Baan et al., 1976). Therefore colicin E3 and cloacin DF13 appear to have a similar action although, owing to the permeability barrier, they have different spectra.

The colicin E3 and cloacin DF13 effects in single step systems, as indicated above, are very weak compared to their strong inhibitory effects in integrated systems for translation. It appears that each round of aminoacyl-tRNA binding suffers 50% inhibition by cloacin DF13 and therefore after 3-6 rounds of elongation of the polypeptide chain, there is practically no synthesis by cloacin DF13-treated ribosomes (Baan et al., 1978a). Furthermore, cloacin DF13 treated ribosomes are unimpaired in the rate or total extent of some reactions involved in initiation such as f-Met-$tRNA_F$ binding, functions of IF-1 and recycling of IF-2 (Baan et al., 1978b).

It seems therefore that as a consequence of splitting the 3' terminal end of the 16S RNA, colicin E3- and cloacin DF13-ribosomes are damaged in such a manner that no reaction is totally blocked but a number of reactions in translation are partly affected.

Chapter 4

Termination

1. The Mechanism of Termination

The termination phase of translation begins when a chain-terminating codon of the mRNA (nonsense codon) (either UAA, UAG or UGA) is recognized, thus determining the interaction of the release factors with the ribosome. A single release factor (RF) has been isolated from eukaryotic systems, GTP being required in the termination step (g_1). On the other hand three release factors (RF-1, RF-2, and RF-3) have been obtained from bacterial systems; release factor RF-1 is active by itself, whereas the activity of RF-2 is strongly potentiated by RF-3. It is most likely, but not definitively proven, that GTP is required in step (g_1) in bacterial systems.

After recognition of the release factor(s) the peptidyl transferase of the larger ribosomal subunit exercises its peptidyl hydrolase activity and cleaves the bond between the peptidyl and the tRNA moieties of peptidyl-tRNA so that the peptide chain is released [step (g_2)]. A further reaction is required for termination to be completed in which the uncharged tRNA is released [step (h)]. This step is not yet resolved in eukaryotic systems but studies in bacterial systems have shown that the ribosomal release factor RR and the elongation factor EF-G are required in a terminating reaction in which GTP hydrolysis takes place.

2. Inhibitors of Termination

2.1 Negamycin

The antibacterial antibiotic negamycin is bactericidal. A simple compound of unusual amino acid structure (Kondo et al., 1971) (Fig. 60), it is very active on gram-positive and gram-negative bacteria and especially interesting because of its activity on Pseudomonas. The antibiotic inhibits protein synthesis by bacteria and bacterial cell-free systems (Mizuno, Nitta and Umezawa, 1970a, 1970b); negamycin also inhibits poly-phenylalanine synthesis by rat liver extracts to a certain extent (Mizuno, Nitta and Umezawa, 1970b). Negamycin has been reported to inhibit initiation (Mizuno, Nitta and Umezawa, 1970b), elongation (Uehara et al., 1972), and termination (Uehara, Hori and Umezawa, 1976b). Indeed the antibiotic inhibits MS2 RNA-directed binding of f-Met-tRNA to ribosomes (Mizuno, Nitta and Umezawa, 1970b), causes a strong misreading on synthetic polyribonucleotide-directed synthesis of polypeptides (Uehara et al., 1972), and further studies suggested that negamycin might be mainly a termination inhibitor (Uehara, Hori and Umezawa, 1974). Thus phage f_2 RNA-directed protein synthesis is not totally inhibited by negamycin even at high concentrations, and a large polypeptide accumulates on the ribosomes, its molecular weight being very close to that of the coat protein although with a higher histidine content. It was therefore proposed that negamycin blocks the termination

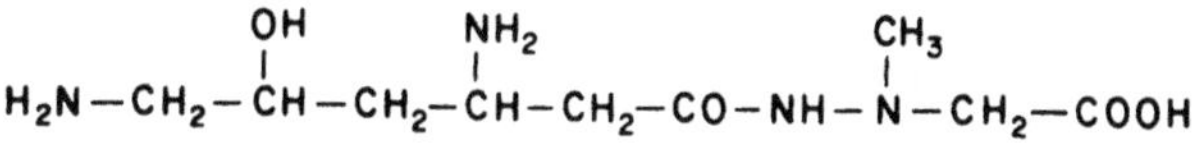

$C_9H_{20}O_4N_4$; MW 248.28

Fig. 60. Chemical structure of negamycin

mechanisms by an unknown mode of action; a higher concentration of the antibiotic is required for the misreading effect (Uehara, Hori and Umezawa, 1976a). Negamycin prevents polysome breakdown, unlike inhibitors of initiation, and has a bactericidal effect even in the presence of chloramphenicol, unlike the aminoglycoside antibiotics which cause misreading (Uehara, Hori and Umezawa, 1976b).

2.2 Other Inhibitors of Termination

Besides the possibility of negamycin, no other case is known of protein synthesis inhibitors blocking translation specifically in the termination phase. A number of the inhibitors described above, however, inhibit termination as a consequence of their interaction with the ribosome, besides inhibiting other steps of translation also. Thus streptomycin and tetracycline block the termination phase of protein synthesis by bacterial ribosomes by preventing release factor(s)-dependent codon recognition [step (g_1)] (Scolnick et al., 1968; Tompkins, Scolnick and Caskey, 1970; Caskey et al., 1971). Similarly, thiostrepton acts as an inhibitor of termination in bacterial systems probably by preventing the interaction of the release factor(s) with the ribosome (Brot et al., 1973). Furthermore it is most likely that some other inhibitors of step (d) are also inhibitors of termination, with the selectivity indicated in Figures 1 and 2.

Peptidyl transferase is involved not only in peptide bond formation [step (e)] but also in the peptidyl-tRNA hydrolysis reaction [step (g_2)]. In fact all the peptidyl transferase inhibitors which have been tested block termination in certain bacterial (Caskey and Beaudet, 1972; Scolnick et al., 1968;

Tompkins, Scolnick and Caskey, 1970; Caskey et al., 1971; Capecchi and Klein, 1969; Menninger, 1971) and mammalian (Caskey et al., 1971; Menninger, 1971; Beaudet and Caskey, 1971) systems with the same specificity shown for peptide bond formation (Figs. 1 and 2). There are however some interesting exceptions. Thus anisomycin inhibits the release of f-Met in a termination assay in the presence of release factor R (Beaudet and Caskey, 1971). However, in a simplified system where f-Met is released in the absence of release factor R, but in the presence of 30% acetone, anisomycin considerably enhances the reaction (Caskey et al., 1971), although under these experimental conditions the drug has a low binding affinity for ribosomes (Barbacid and Vâzquez, 1974a). This system is, of course, artificial, but nevertheless the lack of inhibitory response may be meaningful (Caskey et al., 1971) since a similar finding has also been observed in bacterial systems in the presence of the peptidyl transferase inhibitor lincomycin (Caskey and Beaudet, 1972).

As indicated above, EF-G and GTP appear to be required in step (h) in bacterial systems. It is therefore likely that inhibitors interfering with EF-G activity, such as fusidic acid and certain analogs of GTP, inhibit step (h) in bacteria. Moreover, if a similar mechanism of termination takes place in eukaryotes in step (h), it is to be expected that not only fusidic acid but also other compounds that interfere with EF-2 function and (or) GTP hydrolysis (such as abrin, ricin, PAP, alpha sarcin, diphtheria toxin, and cycloheximide) might also inhibit step (h) in eukaryotic cells. Indeed, cycloheximide has been reported to block termination (Godchaux, Adamson and Herbert, 1967; Rajalakshmi et al., 1971; Oleinick, 1977).

Chapter 5
Miscellaneous Inhibitors of Translation

There are a number of translation inhibitors whose site of action is not yet well known. Thus the antibacterial antibiotic avilamycin blocks protein synthesis in intact bacteria and bacterial cell-free systems (Wolf, 1973). The effect of the antibiotic is located on the 30S ribosomal subunit and inhibits polynucleotide-directed binding of f-Met-tRNA and Phe-tRNA to the ribosome. Since the antibiotic has not been studied in more resolved systems it is not known whether it can be considered an inhibitor of step (d) or of initiation.

The antibacterial antibiotics berninamycin and rubradirin are very active in gram-positive bacteria and in cell-free systems of polypeptide synthesis directed by synthetic polynucleotide (Reusser, 1969; Reusser, 1973). However, berninamycin is a very poor inhibitor of endogenous and phage f_2 RNA-directed synthesis of protein and the site of action of the antibiotic is not yet known (Reusser, 1969). On the other hand rubradirin is an effective inhibitor of phage f_2 RNA-directed synthesis of protein but its site of action is not known (Reusser, 1973).

The streptothricins are strong inhibitors of poly(U)-directed synthesis of polyphenylalanine by E. coli ribosomes. However, the products of mild acid-treatment of the streptothricins show an effect on polyphenylalanine similar to that of the parent antibiotics; nevertheless these products are devoid of antibacterial activity (Khokhlov and Blinov, 1970).

A toxin produced by Shigella dysenteriae inhibits poly(U)-directed polyphenylalanine synthesis and elongation of polypeptide chains by eukaryotic ribosomes (Thompson et al., 1976).

D-galactosamine-containing glycogen polymers, which accumulate in rats treated with large doses of D-galactosamine, are effective inhibitors of amino acid incorporation by rat liver ribosomes in cell-free systems (Mészáros et al., 1976), probably owing to their interaction with ribosomes, forming heavy aggregates which are inactive in translation.

The reaction product between sodium selenite and glutathione (selenodiglutathione) prevents amino acid incorporation by rat liver polysomes by blocking the EF-2; the blocked EF-2 can be regenerated by reduction with glutathione reductase and NADPH (Vernie et al., 1975).

The treatment of Paramecium tetraaurelia with 4% dimethyl sulphoxide reduces the number of polysomes 3 - 4 fold; protein synthesis is correspondingly reduced. This effect appears to be due to a detachment of mRNA from the polysomes rather than a specific effect on initiation and it is not reproduced either in cell-free systems or other organisms (Reisner and Bucholtz, 1977).

A number of oligonucleotides prepared from extracts of eukaryotic cells have been described as having either an inhibitory or an activator effect on translation, and might be of physiological importance. Probably the best characterized of these oligonucleotides is an inhibitor (MW = 6,000) prepared from high speed-supernatants and ribosomal salt washes of dormant and developing *Artemia salina* embryos, which blocks EF-1-dependent binding of aminoacyl-tRNA. Rich in pyrimidines, this inhibitor is complexed with and inactivated by an activator oligonucleotide (MW = 9,000), which is rich in guanines and found in high speed-supernatants and ribosomal salt washes of developing *Artemia salina* embryos; it was therefore postulated that the appearance of this activator might be responsible for the onset of protein synthesis when development resumes (Lee-Huang et al., 1977).

Treatment of *Tetrahymena pyriformis* with sublethal doses of γ-radiation induces polysome dissociation, suggesting that it might prevent initiation of polypeptide synthesis. However, 60% of the polysomes remain intact after γ-radiation and a further effect on elongation and termination has been proposed (Oleinick, Rustad and Kuncio, 1974).

The effects of the polyamines spermine, spermidine and putrescine have been repeatedly tested on the different steps of protein synthesis in bacterial and eukaryotic systems. The results, which appear to be at variance, greatly depend on the system and specific experimental conditions used, the concentration of polyamine also being critical. Thus most of the steps in translation have been reported to be stimulated by certain concentrations of polyamines, but spermine, spermidine, and putrescine have also been reported to inhibit some steps in translation. These polyamines can certainly replace Mg^{++} in some reactions and their inhibitory effects, which have been described in some cases, might be due to an excess of basic ions in these specific experiments. Indeed polyamines and high Mg^{++} concentrations are known to cause misreading in bacterial systems (Friedman and

Weinstein, 1964). However it is widely accepted that the polyamines are involved in the biosynthesis and correct assembly of the small ribosomal subunits (Algranati and Goldemberg, 1977; Goldemberg and Algranati, 1977; reviews).

A micotoxin synthesized by Penicillium roqueforti, known as PRT, has been shown to block protein synthesis by eukaryotic cells and cell-free systems without affecting aminoacyl-tRNA synthesis (Moulê, Jemmali and Darracq, 1978). Since the inhibitory effect of the toxin can be overcome in cell-free systems by addition of an excess of the supernatant fraction, it is possible that PRT act similarly to ricin and related toxins (see above).

A number of bacteriocins, besides colicin E3 and bacteriocin DF13 that are considered above, have been reported to inhibit protein synthesis although their sites of action have not been studied in detail. Thus pneumocins S6 and S8 have been found to inhibit protein synthesis while DNA and RNA synthesis and respiration continue normally. On the other hand pneumocin G196 inhibits protein and RNA synthesis after a delay of 15 min (Reeves, 1972; review).

Chapter 6

GTP Analogs

The GTP analog guanylyl-methylene-diphosphonate has a methylene group replacing the oxygen between the β and γ phosphorus atoms, thus preventing enzymic cleavage at this position (Fig. 61). The structure of this analog has been designed to test the nature of the reactions involving GTP hydrolysis in protein synthesis. Since the GTP analog was first shown to block polyphenylalanine synthesis (Hershey and Monro, 1966), it has been used repeatedly in studies on protein synthesis in the different steps of the initiation, elongation and termination phases in which GTP hydrolysis is involved. The GTP analog guanylyl-imido-diphosphate has also been synthesized to study the role of GTP in translocation (Eckstein, Kettler and Parmeggiani, 1971). Studies with GDP and GTP analogs have concluded that a common or overlapping ribosomal site is involved in the interaction of protein translation factors required for GTP hydrolysis (Modolell and Vázquez, 1975, review). Methylene-diphosphonate and guanylyl-imido-diphosphate can replace GTP in the interaction with the different protein factors and subsequently with the ribosome (Lucas-Lenard and Lipmann, 1971; Bosch, 1972; Haselkorn and Rothman-Denes, 1973; Jackson, 1975; Tate and Caskey, 1974; Modolell and Vázquez, 1975; reviews). However, at least in the cases on which detailed studies were performed (e.g., the elongation cycle), the GTP analogs block or slow down the reactions immediately following the steps in which GTP hydrolysis takes place (Fig. 61). Thus polypeptide chain elongation requires the participation of firstly GTP, secondly elongation factors EF-Ts and EF-Tu (or EF-1), which are involved in the binding of aminoacyl-tRNA, and thirdly EF-G (or EF-2), which is required for translocation of peptidyl-tRNA (Figs. 1 and 2). One molecule of GTP is hydrolyzed during each of these processes (Modolell, Cabrer and Vázquez, 1973b; Cabrer et al., 1976), the hydrolytic reaction requiring the participation of the ribosome and either EF-Tu (or EF-1) plus aminoacyl-tRNA or EF-G (or EF-2). It has been assumed that the energy liberated in the hydrolysis of GTP was used, during aminoacyl-tRNA interaction, to bind this substrate in the correct position to the ribosomal A-site and, during translocation, to assist in rearranging tRNA, mRNA, and ribosomal subunits and thus causing mRNA and peptidyl-tRNA to move. However, guanylyl-methylene-diphosphonate and guanylyl-imido-diphosphate can under some conditions replace GTP in translocation (Inoue-Yokosawa, Ishikawa and Kaziro, 1974; Belitsina, Glukhova and Spirin, 1975; Modolell, Girbés and Vázquez, 1975). Moreover, guanylyl-methylene-diphosphonate promotes

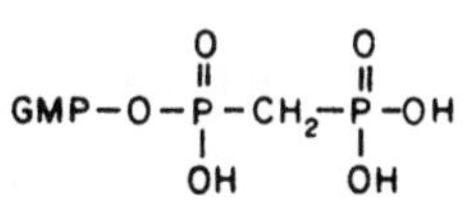

GMP - PCP (synonyms GDPCP and GMP-P-CH_2-P)

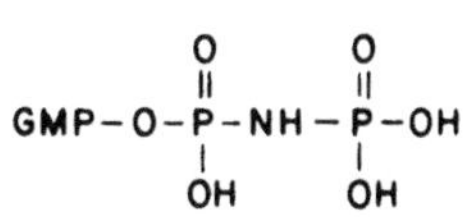

GMP- PNP (synonyms GDPCP and GMP-P-NH-P)

Fig. 61. The GTP analogs

binding of aminoacyl-tRNA (Lucas-Lenard and Lipmann, 1971; Bosch, 1972; Haselkorn and Rothman-Denes, 1973; Jackson, 1975; Tate and Caskey, 1974; Modolell and Vázquez, 1975; reviews) but EF-Tu (or EF-1) remains bound to the ribosome and the aminoacyl-tRNA cannot function as an acceptor substrate for peptide bond formation; however, removal of the EF-Tu (or EF-1) bound by zonal centrifugation relieves this impediment (Yokosawa et al., 1973, 1975). Furthermore guanylyl-iodo-diphosphonate promotes polypeptide synthesis but at a slower rate than GTP (Girbés, Vázquez and Modolell, 1976). Experimental evidence suggests that during elongation of polypeptide chain GTP analogs, EF-Tu, EF-1, EF-G and EF-2 are slowly released from the ribosome and consequently the steps subsequent to aminoacyl binding and translocation, that is, peptide bond formation and aminoacyl-tRNA binding respectively, are delayed. Thus during the elongation cycle GTP hydrolysis is probably essential for fast release of the factors from the ribosome (Girbés, Vázquez and Modolell, 1976; Belitsina, Glukhova and Spirin, 1975; Kaziro, 1973; Möller, 1974). In all likelihood a similar situation occurs in the blockage of initiation and termination when GTP is replaced by its non-hydrolyzable analogs (Lucas-Lenard and Lipmann, 1971; Bosch, 1972; Haselkorn and Rothman-Denes, 1973; Jackson, 1975; Tate and Caskey, 1974; Modolell and Vázquez, 1975; reviews). Thus, in the initiation reaction, when f-Met-$tRNA_F$ is attached to the 70S ribosome in the presence of guanylyl-methylene-diphosphonate, the substrate cannot react with puromycin because the IF-2 guanylyl-methylene-diphosphonate complex remains bound to the ribosome; the substrate is reactive though if this complex is released from the ribosome

(Benne and Voorma, 1972). GTP hydrolysis therefore appears to be required for the fast release of IF-2 (Benne and Voorma, 1972).

A number of other GTP analogs have also been developed to inhibit translation by competing with GTP on the ribosome, although the studies on their mode of action do not reach such satisfactory conclusions as in the case of the GTP analogs described above. Thus guanosine-5'-phosphohypophosphate competes with GTP and inhibits polyphenylalanine synthesis (Remy et al., 1970). The GTP analogs 6-mercaptoguanosine 5'-triphosphate, N^2-methylguanosine 5'-triphosphate, N^2-dimethylguanosine 5'-triphosphate,and 8-bromoguanosine 5'-triphosphate are also inhibitors of translation (Uno et al., 1971).

Chapter 7

Selectivity and Specificity Reconsidered

Despite the broad scope of the inhibitors of translation mentioned above and presented in Tables 2-4, there are some cases in which certain antibiotics appear to have narrower spectra of selectivity. Thus lincomycin is very active on ribosomes from gram-positive bacteria and moderately active on blue-green algae ribosomes, but has low activity on ribosomes from gram-negative bacteria in preventing peptide bond formation (Chang and Weisblum, 1967; Celma, 1971; Monro et al., 1971). Furthermore, tenuazonic acid inhibits protein synthesis by mammalian but not by yeast ribosomes (Carrasco and Vázquez, 1973b), cycloheximide is inactive on ribosomes of wild type *Saccharomyces fragilis* (Rao and Grollman, 1967) and *Saccharomyces lactis*, while fusidic acid is inactive on ribosomes from *Neurospora crassa* mitochondria (Grandi, Helma and Kuntzel, 1971) and in protein synthesizing systems from sporulating *Bacillus subtilis* (Fortnagel and Bergman, 1973). Furthermore, normal ribosomes, sensitive to fusidic acid, are required for the initiation of normal sporulation, but ribosomes functioning throughout the development period after septum formation do not form the fusidic acid-EF-G·GTP·ribosome complex and are therefore resistant to fusidic acid (Fortnagel and Freese, 1977). Fusidic acid-resistant, sporulation-defective mutants have been isolated and shown that the mutation resulted from an alteration in elongation factor EF-G (Kobayashi, Kobayashi and Kobayashi, 1977).

It has also been reported that erythromycin and lincomycin do not inhibit protein synthesis by ribosomes from rat liver mitochondria (Firkin and Linnane, 1969; Towers et al., 1972; Towers, Kellerman and Linnane, 1973), although this proposal is at variance with some other reports (Kroon and DeVries, 1970, 1971; Ibrahim and Beattie, 1973; Ibrahim, Burke and Beattie, 1974). The inhibitory effect of 7-methylguanosine-5'-monophosphate is restricted to systems with the sequence $m^7G^{5'}$ppp at the 5' end of the mRNA (Roman et al., 1976; Hickey,Weber and Baglioni, 1976; Shafritz et al., 1976). Moreover, the cell permeability barrier results in a narrower spectra of inhibitory activity than might be expected from Tables 2-4. Thus it is well known that owing to a cell permeability barrier lincomycin, streptogramins, macrolides, and some other antibiotics are inactive on gram-negative bacteria, most antibacterial antibiotics are not active in Mycobacteria,and most inhibitors which act on eukaryotic ribosomes are inactive in intact yeast. Interestingly, it has recently been shown that various translation inhibitors are inactive in tissue culture cells (Contreras, Vázquez and Carrasco, 1978), but the permeability barrier is altered in picornavirus-infected cells,resulting in an inhibitory effect of the inhibitors in the infected cells (Contreras and Carrasco, 1978).

On the other hand, a wider spectrum of activity than that indicated in Tables 2-4 has been observed in a few cases. Thus sensitivity of brain mitochondrial ribosomes to emetine has been reported (Ibrahim, Burke and Beattie, 1974; Lietman, 1971) and a yeast mutant with cytoplasmic 80S-type ribosomes sensitive to streptomycin has also been described (Bayliss and Vinopal, 1971; Bayliss and Ingraham, 1974). Besides the mitochondrial system, another protein synthesizing system which is as sensitive to chloramphenicol as the mitochondrial ribosomes has been reported to function in synaptosomes (Ramírez, Levitan and Mushynski, 1972).

Griseoviridin and streptogramin A are antibacterial antibiotics, but at rather high concentrations are also active on eukaryotic ribosomes in cell-free systems (Barbacid, Contreras and Vázquez, 1975). Aminoglycoside antibiotics containing the paromamine moiety (lividomycin B, paromomycin, gentamicin A, and kanamycin C) inhibit translation by eukaryotic ribosomes and cause misreading at fairly low concentrations. Neamine, gentamicin B, gentamicin C1A, kanamycin A, and kanamycin 1A also cause misreading by eukaryotic ribosomes but only at rather high concentrations (Palmer and Wilhelm, 1978; Wilhelm, Pettit and Jessop, 1978). These few examples illustrate some limited but important differences in the sensitivity of ribosomes to inhibitors which are not expressed in Table 1. Furthermore, ribosomes, initiation factors, and elongation factors might play an important role in the translational control of protein synthesis in eukaryotic and transformed cells and it is likely that sensitivity to inhibitors of protein synthesis depends on the state or phase of the cells; inhibitors of protein synthesis might therefore be important tools in the study of ribosome diversity and control of translation in higher cells.

Although most of the translation inhibitors indicated in Tables 2-4 are considered to act by specifically preventing protein synthesis, there are a number of cases in which their action is less specific. Thus, in intact cells, edeine A1 is mainly an inhibitor of DNA synthesis,and cycloheximide can inhibit not only translation but also DNA and RNA synthesis. Furthermore aminoglycoside antibiotics are known to have many interactions in cell-free systems which might result in a number of effects on intact cells other than protein synthesis inhibition (Brock, 1966; review). Moreover, even chloramphenicol, which appears to have a clear and specific action in bacteria, has been shown to inhibit DNA synthesis at rather high concentrations in eukaryotic cells (Freeman, Patel and Haldar, 1977); this effect appears to be the main reason for the irreversible toxicity of the antibiotic in certain cases in medical use (various contributions in Postgraduate Medical Journal, 1974).

The above considerations should be taken into account when using translation inhibitors as biochemical tools in intact cells. Obviously the risks of erroneous interpretations are smaller when the inhibitors are used in cell-free systems to study specific reactions.

References

ABRAHAM, A.P., HEATLEY, N.G., BROOKES, P., FULLER, A.T., WALKER, J. (1956) Probable identity of antibiotic produced by a spore-bearing bacillus of the B. pumilus group with micrococcin. Nature 178, 44-45

ABREU, S.L., LUCAS-LENARD, J. (1976) Cellular protein synthesis shut-off by mengovirus: translation of nonviral and viral mRNA's in extracts from uninfected and infected Ehrlich ascites tumor cells. J. Virol. 18, 182-194

ABREU, S.L., LUCAS-LENARD, J. (1977) Inhibition of cellular and viral protein synthesis by 3-methyleneoxindole. Ant. Agents Chemother. 11, 521-527

AHMED, A. (1968a) Mechanism of inhibition of protein synthesis by spiramycin. Biochim. Biophys. Acta 166, 205-217

AHMED, A. (1968b) Altered ribosomes in spiramycin-resistant mutants of Bacillus subtillis. Biochim. Biophys. Acta 166, 218-228.

AKITA, E., TSURUOKA, T., EZAKI, N., NIIDA, T. (1970) Studies on antibiotic SF-733, a new antibiotic. II. Chemical structure of antibiotic SF-733. J. Antibiotics 23, 173-183

ALEXANDER, D.C., JORDAN, D.C., McKAGUE, M. (1969) Inhibition by viomycin of cell-free protein synthesis in sensitive and resistant strains of Rhizobium meliloti. Can. J. Microbiol. 47, 1092-1094

ALGRANATI, D., GOLDEMBERG, S.H. (1977) Polyamines and their role in protein synthesis. Trends in Biochemical Research 2, 272-274

ANDERSON, B., HODGKIN, D.C., VISWAMITRA, M.A. (1970) The structure of thiostrepton. Nature 225, 233-235

ANDERSON, L.A., SMILLIE, R.M. (1966) Binding of chloramphenicol by ribosomes from chloroplasts. Biochem. Biophys. Res. Commun. 23, 535-539

ANDERSON, W.F., BOSCH, L., COHN, W.E., LODISH, H., MERRICK, W.C., WEISSBACH, H., WITTMANN, H.G., WOOL, I.G. (1977) International Symposium on protein synthesis. Summary of Fogarty Center-NIH Workshop held in Bethesda, Maryland on 18-20 October, 1976. FEBS Letters 76, 1-10

ARGOUDELIS, A.D., BRODASKY, T.F. (1972) Studies with streptomyces caelestis. I. New celesticetins. J. Antibiotics 25, 194-196

ARIATTI, M., HAWTREY, A.O. (1975) Synthesis of cyclohexyl-puromycin and its reaction with N-acetylphenylalanyl-transfer ribonucleic acid on rat liver ribosomes. Biochem. J. 145, 169-175

ASATRYAN, L.S., SPIRIN, A.S. (1975) Non-enzymatic translocation in ribosomes from streptomycin-resistant mutants of Escherichia coli. Mol. Gen. Genetics 138, 315-321

AYUSO, M., GOLDBERG, I.H. Pactamycin inhibition of eukaryotic polypeptide synthesis dependent on added initiation factors. Biochim. Biophys. Acta 294, 118-122

BAAN, R.A., DUIJFJES, J.J., van LEERDAM, E., van KNIPPENBERG, P.H., BOSCH, L. (1976) Specific in situ cleavage of 16S ribosomal RNA of Escherichia coli interferes with the function of initiation factor IF-1. Proc. Nat. Acad. Sci. 73, 702-706

BAAN, R.A., FRIJMANN, M., van KNIPPENBERG, P.H. BOSCH, L. (1978a) Consequences of a specific in situ cleavage of 16S ribosomal RNA for polypeptide chain elongation. Eur. J. Biochem. 87, 131-136

BAAN, R.A., NAAKTGEBOREN, N., van CHARLDORP, R., van KNIPPENBERG, P.H., BOSCH, L. (1978b) Consequences of a specific in situ cleavage of 16S ribosomal RNA for polypeptide chain initiation. Eur. J. Biochem. 87, 137-142

BAASKE, D.M., HEINSTEIN, P. (1977) Cytotoxicity and cell cycle specificity of homoharringtonine. Ant. Agents Chemother. 12, 298-300

BABLANIAN, R. (1972) Depression of macromolecular synthesis in cells infected with guanidine-dependent polyvirus under restrictive conditions. Virology 47, 255-259

BAEZ, A., ANGOSO, M., ALONSO, G., VÂZQUEZ, D. (1977) Labelling of narciclasine by tritium exchange. Biochemie 59, 751-753

BAEZ, A., VAZQUEZ, D. (1978) Binding of (^{3}H) narciclasine to eukaryotic ribosomes. A study on a structures activity relationship. Biochim. Biophys. Acta 518, 95-103

BAGLIONI, C., JACOBS-LORENA, M., MEADE, H. (1972) The site of action of inhibitors of initiation of protein synthesis in reticulocytes. Biochim. Biophys. Acta 277, 188-197

BAKER, B.R., SCHAUB, R.E., JOSEPH, J.P., WILLIAMS, J.H. (1955) Puromycin. Synthetic studies. IX. Total synthesis. J. Amer. Chem. Soc. 77, 12-15

BALIGA, B.S., COHEN, S.A., MUNRO, H.N. (1970) Effect of cycloheximide on the reaction of puromycin with polysomebound peptidyl-tRNA. FEBS Letters 8, 249-252

BALIGA, B.S., PRONCZUCK, A.W., MUNRO, H.N. (1969) Mechanism of cycloheximide inhibition of protein synthesis in a cell-free system prepared from rat liver. J. Biol. Chem. 244, 4480-4489

BALIGA, B.S., ZAHRINGER, J., TRACHTENBERG, M., MOSKOWITZ, M.A., MUNRO, H.N. (1976) Mechanism of D-amphetamine inhibition of protein synthesis. Biochim. Biophys. Acta 442, 239-250

BALLESTA, J.P.G., MONTEJO, V., HERNANDEZ, F., VAZQUEZ, D. (1974) Alteration of ribosomal proteins and functions by 2-methoxy-5-nitrotropone. Eur. J. Biochem. 42, 167-175

BALLESTA, J.P.G., VAZQUEZ, D. (1972a) Reconstitution of the 50S ribosome subunit. Role of proteins L7 and L12 in the GTPase activities. Site of action of thiostrepton. FEBS Letters 28, 337-342

BALLESTA, J.P.G., VAZQUEZ, D. (1972b) Elongation factor T-dependent hydrolysis of guanosine triphosphate resistant to thiostrepton. Proc. Nat. Acad. Sci. 69, 3058-3062

BAMBURG, J.R. (1976) Chemical and biochemical studies of the trichothecene mycotoxins. In: Micotoxins and Other Fungal Related Food Products (Rodricks, J.V., ed.), pp. 144-162. American Chemical Society, Washington, D.C. USA

BARBACID, M., CONTRERAS, A., VAZQUEZ, D. (1975) The mode of action of griseoviridin at the ribosome level. Biochim. Biophys. Acta 395, 347-354

BARBACID, M., FRESNO, M., VAZQUEZ, D. (1975) Inhibitors of polypeptide elongation on yeast polysomes. J. Antibiotics 28, 453-462

BARBACID, M., VAZQUEZ, D. (1973) Preparation of pure (^{3}H) anisomycin by tritium exchange labeling. Anal. Biochem. 56, 16-25

BARBACID, M., VAZQUEZ, D. (1974a) (^{3}H) Anisomycin binding to eukaryotic ribosomes. J. Mol. Biol. 84, 603-623

BARBACID, M., VAZQUEZ, D. (1974b) Binding of (acetyl-^{14}C)-trichodermin to the peptidyl transferase centre of eukaryotic ribosomes. Eur. J. Biochem. 44, 437-444

BARBACID, M., VAZQUEZ, D. (1974c) (G-^{3}H)Gougerotin binding to ribosomes. Heterogeneity of eukaryotic ribosomes. Eur.J. Biochem. 44, 445-453

BARBACID, M., VAZQUEZ, D. (1975) Ribosome changes during translation. J.Mol.Biol. 93, 449-463

BARZA, M., BROWN, R.B., SHANKS, C., GAMBLE, C., WEINSTEIN, L. (1975) Relation between lipophilicity and pharmacological behavior of minocycline, doxycycline, tetracycline and oxytetracycline in dogs. Ant.Agents Chemother. 8, 713-720

BATTANER, E., VAZQUEZ, D. (1971) Inhibitors of protein synthesis by ribosomes of the 80S type. Biochim.Biophys.Acta 154, 316-330

BAXTER, R., McGOWAN, J.E. (1976) MDMP action: Degradative effects on polyribosomes from wheat roots and the inhibition of protein synthesis initiation. J.Exptl.Botany 27, 525-531

BAYLISS, F.T., INGRAHAM, J.L. (1974) Mutation in Saccharomyces cerevisiae conferring streptomycin and cold sensitivity by affecting ribosome formation and function. J.Bacteriol. 174, 1339-1341

BAYLISS, F.T., VINOPAL, R.T. (1971) Selection of ribosomal mutants by antibiotic suppresion in yeast. Science 174, 1339-1341

BAZZICALUPO, M., PARISI, B., PIRALI, G., POLSINELLI, M., SALA, F. (1975) Chemical and biochemical characterization of a ribosomal mutant of Bacillus subtilis resistant to sporangiomycin. Ant.Agents Chemother. 8, 651-656

BEAUDET, A.L., CASKEY, C.T. (1971) Mammalian peptide chain termination. II. Codon specificity and GTPase activity of release factor. Proc.Nat.Acad.Sci. 68, 619-624

BEGUERET, J., PERROT, M., CROUZET, M. (1977) Ribosomal proteins in the fungus Podospora anserina: evidence for an electrophoretically altered 60S protein in a cycloheximide resistant mutant. Mol.Gen.Genetics 156, 141-144

BELITSINA, N.V., GLUKHOVA, M.A., SPIRIN, A.S. (1975) Translocation in ribosomes by attachment-detachment of elongation factor G without GTP cleavage: evidence from a collumn-bound ribosome system. FEBS Letters 54, 35-38

BENNE, R., VOORMA, H.O. (1972) Entry site of formylmethionyl-tRNA. FEBS Letters 20, 347-351

BENNETT, L.L., WARD, V.L., BROCKMAN, R.W. (1965) Inhibition of protein synthesis in vitro by cycloheximide and related glutarimide antibiotics. Biochim.Biophys.Acta 103, 478-485

BENSON, S.C. (1977) Correlation of inhibition of ribosomal GTPase and protein synthesis by abrin and ricin. Biochem.Biophys. Res.Commun. 75, 845-850

BENSON, S., OLSNES, S., PIHL, A., SKORVE, J., ABRAHAM, K.A. (1975) On the mechanism of protein synthesis inhibition by abrin and ricin. Inhibition of the GTP-hydrolysis. Site on the 60S ribosomal subunit. Eur.J.Biochem. 59, 573-580

BENVENISTE, R., DAVIES, J. (1973a) Mechanism of antibiotic resistance in bacteria. Ann.Rev.Biochem. 42, 471-506

BENVENISTE, R., DAVIES, J. (1973b) Structure-activity relationships among the aminoglycoside antibiotics: role of hydroxyl and amino groups. Ant.Agents Chemother. 4, 402-409

BENZER, S., CHAMPE, S.P. (1962) A change of nonsense to sense in the genetic code. Proc.Nat.Acad.Sci. 48, 1114-1121

BERES, L., LUCAS-LENARD, J. (1975) Studies on the in vitro synthesis of ppGpp and pppGpp. Biochim.Biophys.Acta 395, 80-90

BERGER, J., LEHR, H.H., TEITEL, S., MAEHR, H., GRUNDBERG, E. (1973) A new antibiotic X-5108 of Streptomyces origin. I. Production, isolation and properties. J.Antibiotics 26, 15-22

BERLINGUET, L., NORMAND, A. (1968) Effects of polyamino acids, histones and heparin on amino acid incorporation into ribosomal proteins from rat liver. Biochim.Biophys.Acta 161, 509-517

BERMEK, E. (1972) Formation of a complex involving ADP-ribosylated human translocation factor, guanosine nucleotide and ribosomes. FEBS Letters 23, 95-99

BERMEK, E. (1976) Interactions of adenosine diphosphate-ribosylated elongation factor 2 with ribosomes. J.Biol.Chem. 251, 6544-6549

BERMEK, E., KRAMER, W., MONKEMEYER, H., MATTHAEI, H. (1970) Mechanisms in protein synthesis XII. Sites of action of showdomycin in a cell-free polyphenylalanine synthesizing systems from human lymphatic tissue: Ribosomes and elongation factor IF II. Biochem.Biophys.Res.Commun. 40, 1311-1318

BERMEK, E., MATTHAEI, H. (1970) The effect of antibiotics on an optimyzed polyphenylalanine synthesizing cell-free system from human lymphatic tissue. Hoppe Seyler Z.Physiol.Chem. 351, 1377-1383

BERMEK, E., MONKEMEYER, H., BERG, R. (1971) The role of SH-groups of human ribosomal subunits in polypeptide synthesis. Biochem.Biophys.Res.Commun. 45, 1294-1299

BERNABEU, C., VAZQUEZ, D., BALLESTA, J.P.G. (1976) Activities of protein-deficient particles derived from 50S ribosomal subunits by NH_4Cl/ethanol treatment. Eur.J.Biochem. 69, 233-241

BERNABEU, C., VAZQUEZ, D., BALLESTA, J.P.G. (1977) The involvement of protein L16 on ribosomal peptidyl transferase activity. Eur.J.Biochem. 79- 469-472

BERNARDI, A., LEDER, P. (1970) Protein biosynthesis in Escherichia coli. Purification and characterization of a mutant G factor. J.Biol.Chem. 245, 4263-4268

BETHUNE, M.P.de., NIERHAUS, K.H. (1978) Characterisation of the binding of virginiamycin S to Escherichia coli ribosomes. Eur.J.Biochem. 86, 187-191

BHUYAN, B.K., DIETZ, A., SMITH, C.G. (1961) Pactamycin, a new antitumor antibiotic. I. Discovery and biological properties. Ant.Agents Chemother. pp. 184-190

BIRCH, A.J., HOLZAPFEL, C.W., RICKARDS, R.W. (1966) The structure and some aspects of the biosynthesis of pleuromutilin. Tetrahedron Suppl. 8, 359-387

BIRGE, E.A., KURLAND, C.G. (1969) Altered ribosomal protein in streptomycin dependent Escherichia coli ribosomes. Science 166, 1282-1284

BIRGE, E.A., KURLAND, C.G. (1970) Reversion of a streptomycin-dependent strain of Escherichia coli. Science 166, 1282

BISHOP, J.O. (1968) Effect of puromycin and sodium fluoride on reticulocyte ribosomal monomers and subribosomal particles. Arch.Biochem.Biophys. 125, 449-451

BISWAS, D.K., GORINI, L. (1972)The attachment site of streptomycin to the 30S ribosomal subunit. Proc.Nat.Acad.Sci. 69, 2141-2144

BLAT, C., LOEB, J.E. (1971) Effect of glucanol phosphorylation of some rat liver ribosomal proteins in vivo. FEBS Letters 18, 124-126

BLOCH, A., COUTSOGEORGOPOULOS, C. (1966) Inhibition of protein synthesis by amicetin, a nucleoside antibiotic. Biochemistry 5, 3345-3351

BLUMENTHAL, T., DOUGLASS, J., SMITH, D. (1977) Conformational alteration of protein synthesis elongation factor EF-Tu by EF-Ts and kirromycin. Proc.Nat.Acad.Sci. 74, 3264-3267

BODANSZKY, M., DUTCHER, J.D., WILLIAMS, N.J. (1963) The establishment of the identity of thiostrepton with thiactin (bryamycin). J.Antibiotics 16, 76-79

BODLEY, J.W., GODFREDSEN, W.O. (1972) Studies on translocation XI: Structure-function relationships of the fusidane-type antibiotics. Biochem.Biophys.Res.Commun. 46, 871-877

BODLEY, J.W., LIN, L. (1970) Interaction of E. coli G factor with the 50S ribosomal subunit. Nature 227, 60-62

BODLEY, J.W., LIN, L., HIGHLAND, J.H. (1970) Studies on translocation. IV. Thiostrepton prevents the formation of a ribosome-G factor-guanine nucleotide complex. Biochem. Biophys.Res.Commun. 41, 1406-1411

BODLEY, J.W., LIN, L., SALAS, M.L., TAO, M. (1970) Studies on translocation V: Fusidic acid stabilization of a eukaryotic ribosome-translocation factor-GDP complex. FEBS Letters 11, 153-156

BODLEY, J.W., ZIEVE, F.J., LIN, L., ZIEVE, S.T. (1969) Formation of the ribosome-G factor-GDP complex in the presence of fusidic acid. Biochem.Biophys.Res.Commun. 37, 437-443

BOLLEN, A., CABEZON, T., WILDE, M.de., VILLARROEL, R., HERZOG, A. (1975) Alterations of ribosomal protein S17 by mutation linked to neamine resistance in Escherichia coli. J.Mol.Biol. 99, 795-806

BOLLEN, A., DAVIES, J., OZAKI, M., MIZUSHIMA, S. (1969) Ribosomal protein conferring sensitivity to the antibiotic spectinomycin in Escherichia coli. Science 165, 85-86

BOON, T. (1971) Inactivation of ribosomes in vitro by colicin E3. Proc.Nat.Acad.Sci. 68, 2421-2425

BOON, T. (1972) Inactivation of ribosomes in vitro by colicin E3 and its mechanism of action. Proc.Nat.Acad.Sci. 69, 549-552

BOSCH, L. (ed.) (1972) In "The Mechanism of Protein Synthesis and its Regulation", Amsterdam: North-Holland Publishing Co

BOTH, G.W., FURUICHI, Y., MUTHUKRISHNAN, S., SHATKIN, A.J. (1975) Ribosome binding to reovirus mRNA in protein synthesis requires 5' terminal 7-methylguanosine. Cell 6, 185-195

BOTTENSTEIN, J.E., VELLIS, J.de. (1976) Divalent cation ionophore A23187: A potent protein synthesis inhibitor. Biochem.Biophys. Res.Commun. 73, 486-493

BOWMAN, C.M. (1972) Inactivation of ribosomes by colicin E3 in vitro: requirement for 50S ribosomal subunits. FEBS Letters 32, 73-75

BOWMAN, C.M., DAHLBERG, J.E., IKEMURA, T., KONISKY, J., NOMURA, M. (1971) Specific inactivation of 16S ribosomal RNA induced by colicin E3 in vivo. Proc.Nat.Acad.Sci. 68, 964-968

BOWMAN, C.M., SIDIKARO, J., NOMURA, M. (1971) Specific inactivation of ribosomes by colicin E3 in vitro and mechanism of immunity in colicinogenic cells. Nature New Biol. 234, 133-137

BRAKIER-GINGRAS, L., LACOSTE, L., BOILEAU, G. (1974) Streptomycin resistance and ribosomal proteins: amino acid substitutions in the str protein of one streptomycin-resistant mutant of Escherichia coli K12 after mutagenesis with ethylmethanesulfonate. Can.J.Biochem. 52, 304-309

BRAKIER-GINGRAS, L., PROVOST, L., DUGAS, H. (1974) Conformational changes induced in Escherichia coli ribosomes by streptomycin. Biochem.Biophys.Res.Commun. 60, 1238-1244

BRAUCHLI, P., DEUFOLEU, V., BUDZIKIEWICZ, H., DJERASSI, C. (1964) The structure of tubulosine, a novel alkaloid from Pegonopus tubulosus (DC) Shumann. J.Amer.Chem.Soc. 86, 1895-1896

BREGA, A., FALASCHI, A., CARLI, L.de., PAVAN, M. (1968) Studies on the mechanism of action of pederin. J.Cell.Biol. 36, 485-496

BREITBART, H., ATLAN, H., ELTES, F., HERZBERG, M. (1975) Interaction between membrane properties and protein synthesis in reticulocytes: a two step inhibition of protein synthesis by valinomycin. Mol.Biol.Reports 2, 167-173

BRENNER, S., BECKWITH, J.R. (1965) Ochre mutants, a new class of suppressible nonsense mutants. J.Mol.Biol. 13, 629-637

BRETSCHER, M.S., MARCKER, K.A. (1966) Polypeptidyl-sribonucleic acid and amino-acyl-sribonucleic acid binding sites on ribosomes. Nature 211, 380-384

BROCK, T.D. (1966) Streptomycin. In "Biochemicals Studies of Antimicrobial Drugs" (Newton, B.A. and Reynilds, P.E., eds.), pp. 131-168, University Press, Cambridge, England

BROT, N., MARCEL, R., YAMASAKI, E., WEISSBACH, H. (1973) Further studies on the role of 50S ribosomal proteins in protein synthesis. J.Biol.Chem. 248, 6952-6956

BROT, N., SPEARS, C., WEISSBACH, H. (1971) The interaction of transfer factor G, ribosomes, and guanosine nucleotides in the presence of fusidic acid. Arch.Biochem.Biophys. 143, 286-296

BROT, N., YAMASAKI,E., REDFIELD, B., WEISSBACH, H. (1972) The properties of an E. coli ribosomal protein required for the function of factor G. Arch.Biochem.Biophys. 148, 148-155

BUCHWALD, I., HACKETT, P.B., EGBERTS, E., TRAUB, P. (1977) 2'(3'),5'-ADP inhibits initiation-dependent protein synthesis in a cell-free system from Ehrlich ascites tumor cells. Mol. Biol.Reports 3, 315-321

BUCKEL, P., BUCHBERGER, A., BOCK, A., WITTMANN, H.G. (1977) Alteration of ribosomal protein L6 in mutants of Escherichia coli resistant to gentamicin. Mol.Gen.Genetics 158, 47-54

BURKE, D.C. (1977) The biological effects of double-stranded RNA-some unanswered questions. TIBS 2, 249-251

BURNS, D.J.W., CUNDLIFFE, E. (1973) Bacterial-protein synthesis. A novel system for studying antibiotic action in vivo. Eur.J. Biochem. 37, 570-574

BURNS, K., CANNON, M., CUNDLIFFE, E. (1974) A resolution of conflicting reports concerning the mode of action of fusidic acid. FEBS Letters 40, 219-223

BUTTERSBY, A.R., TURNER, J.C. (1960) Ipecacuanha alkaloids. Part. V. Stereospecific synthesis of (+)-O-methylpschotrine and (-)-emetine. J.Chem.Soc. 717-725

CABAÑAS, M.J., VAZQUEZ, D., MODOLELL, J. (1978) Dual interference of hygromycin B with ribosomal translocation and with aminoacyl-tRNA recognition. Eur.J Biochem. 87, 21-28

CABRER, B., SAN-MILLAN, M.J., VAZQUEZ, D., MODOLELL, J. (1976) Stoichiometry of polypeptide chain elongation. J.Biol.Chem. 251, 1718-1722

CABRER, B., VAZQUEZ, D., MODOLELL, J. (1972) Inhibition by elongation factor EF G of aminoacyl-tRNA binding to ribosomes. Proc.Nat.Acad.Sci. 69, 733-736

CANNON,M., BURNS, K. (1971) Modes of action of erythromycin and thiostrepton as inhibitors of protein synthesis. FEBS Letters 18, 1-5

CANNON, M., CABEZON, T., BOLLEN, A. (1974) Mapping of neamine resistance: identification of two genetic loci, nea A and nea B. Mol.Gen.Genetics 130, 321-326

CANNON, M., JIMENEZ, A., VAZQUEZ, D. (1976) Competition between trichodermin and several other sesquiterpene antibiotics for binding to their receptor site(s) on eukaryotic ribosomes. Biochem.J. 160, 137-145

CANNON, M., SMITH, K.E., CARTER, C.J. (1976) Prevention, by ribosome-bound nascent polyphenylalanine chains, of the functional interaction of T-2 toxin with its receptor site. Biochem.J. 156, 289-294

CAPECCHI, M.R., KLEIN, H.A. (1969) Characterization of three proteins involved in polypeptide chain termination. Cold Spring Harb. Symp.Quant.Biol. 34, 469-477

CARDANI, C.,GHIRINGHELLI, D., MONDELLI, R., QUILICO, A. (1965) The structure of pederin. Tetrahedron Letters 29, 2537-2545.

CARRASCO, L. (1977) The inhibition of cell functions after viral infection. A proposed general mechanism. FEBS Letters 76, 11-15

CARRASCO, L. (1978) Membrane leakiness after viral infection and a new approach to the development of antiviral agents. Nature 272, 694-699

CARRASCO, L., BARBACID, M., VAZQUEZ, D. (1973) The trichodermin group of antibiotics, inhibitors of peptide bond formation by eukaryotic ribosomes. Biochim.Biophys.Acta 312, 368-376

CARRASCO, L., FERNANDEZ-PUENTES, C., VAZQUEZ, D. (1975) Effects of ricin on the ribosomal sites involved in the interaction of the elongation factors. Eur.J.Biochem. 54, 499-503

CARRASCO, L., FERNANDEZ-PUENTES, C., VAZQUEZ, D. (1976) Antibiotics and compounds affecting translation by eukaryotic ribosomes. Specific enhancement of aminoacyl-tRNA binding by methylxanthines. Mol.Cell.Biochem. 10, 97-122

CARRASCO, L., FRESNO, M., VAZQUEZ, D. (1975) Narciclasine: an antitumour alkaloid which blocks peptide bond formation by eukaryotic ribosomes. FEBS Letters 52, 236-239

CARRASCO, L., JIMENEZ, A., VAZQUEZ, D. (1976) Specific inhibition of translocation by tubulosine in eukaryotic polysomes. Eur.J. Biochem. 64, 1-5

CARRASCO, L., SMITH, A.E. (1976) Sodium ions and the shut-off of host cell protein synthesis by picornaviruses. Nature 264, 807-809

CARRASCO, L., VAZQUEZ, D. (1972) Survey of inhibitors in different steps of protein synthesis by mammalian ribosomes. J.Antibiotics 25, 732-737

CARRASCO, L., VAZQUEZ, D. (1973a) Ribosomal sites involved in binding of aminoacyl-tRNA and EF 2. Mode of action of fusidic acid. FEBS Letters 32, 152-156

CARRASCO, L., VAZQUEZ, D. (1973b) Differences in eukaryotic ribosomes detected by the selective action of an antibiotic. Biochim.Biophys.Acta 319, 209-215

CARTER, C.J., CANNON, M. (1977) Structural requirements for the inhibitory action of 12-13-cpoxytrichothecenes on protein synthesis in eukaryotes. Biochem.J. 166, 399-409

CARTER, C.J., CANNON, M. (1978) Inhibition of eukaryotic ribosomal function by the sesquiterpenoid antibiotic fusarenon-X. Eur.J. Biochem. 84, 103-111

CARTER, C.J., CANNON, M., SMITH, K.E. (1976) Inhibition of protein synthesis in reticulocyte lysates by trichodermin. Biochem.J. 154, 171-178

CARTER, W.A., LEVY, H.B. (1967) Ribosomes: effect on interferon on their interaction with rapidly labeled. Science 155, 1254-1256

CASJENS, S.R., MORRIS, A.J. (1965) The selective inhibition of protein assembly by gougerotin. Biochim.Biophys.Acta 108, 677-686

CASKEY, C., BEAUDET, A. (1972) Antibiotic inhibitors of peptide chain termination. In "Molecular Mechanism of Antibiotic Action on Protein Biosynthesis and Membranes" (Muñoz, E., García-Ferrandiz, F. and Vázquez, D., eds.), pp. 326-336, Elsevier Scientific Publishing Company, Amsterdam-London-New York

CASKEY, C.T., BEAUDET, A.L., SCOLNICK, E.M., ROSMAN, M. (1971) Peptidyl transferase hydrolysis of fMet-tRNA. Proc.Nat.Acad. Sci. 68, 3163-3167

CAWLEY, D.B., HEDBLOM, M.L., HOFFMAN, E.J., HOUSTON, L.L. (1977) Differential ricin sensitivity of rat liver and wheat germ ribosomes in polyuridic acid translation. Arch.Biochem. Biophys. 182, 690-695

CAWTHON, M.L., BITTE, L.F., KRYSTOSEK, A., KABAT, D. (1974) Effect of cyclic adenosine 3': 5'-monophosphate on ribosomal protein phosphorylation in reticulocytes. J.Biol.Chem. 249, 275-278

CELMA, M.L. (1971) Biosíntesis de proteínas en bacterias: fijación de sustratos a ribosomas en la formación de enlace peptídico. Inhibición específica por antibióticos. Ph.D.Thesis. Faculty of Pharmacy. Madrid University. Madrid

CELMA, M.L., EHRENFELD, E. (1974) Effect of poliovirus double-stranded RNA on viral and host-cell protein synthesis. Proc. Nat.Acad.Sci. 71, 2440-2444

CELMA, M.L., MONRO, R.E., VAZQUEZ, D. (1970) Substrate and antibiotic sites at the peptidyl transferase centre of E. coli ribosomes. FEBS Letters 6, 273-277

CELMA, M.L., MONRO, R.E., VAZQUEZ, D. (1971) Substrate and antibiotic binding sites at the peptidyltransferase centre of E. coli ribosomes. FEBS Letters 13, 247-251

CELMA, M.L., VAZQUEZ, D., MODOLELL, J. (1972) Failure of fusidic acid and siomycin to block ribosomes in the pretranslocated state. Biochem.Biophys.Res.Commun. 48, 1240-1246

CERIOTTI, G. (1967) Narciclasine: an antimitotic substance from Narcissus bulbs. Nature 213, 595-596

CERNA, J., JONAK, J., RYCHLIK, I. (1971) Effect of the macrolide antibiotics on the ribosomal peptidyl transferase in cell-free systems derived from Escherichia coli B and erythromycin-resistant mutant of Escherichia coli B. Biochim.Biophys.Acta 240, 109-121

CERNA, J., LICHTENTHALER, F.W., RYCHLIK, I. (1971) The effect of gougerotin analogues on ribosomal peptidyl transferase. FEBS Letters 14, 45-48

CERNA, J., RYCHLIK, I. (1972) The effect of antibiotics on the substrate binding to the acceptor and donor site of ribosomal peptidyltransferase of an erythromycin-resistant mutant of Escherichia coli. Biochim.Biophys.Acta 287, 292-300

CERNA, J., RYCHLIK, I., FRAYEVSKY, A.A., GOTTIKH, B.P. (1976) Effect of CMP and its analogson peptidyl transferase activity. In "Ribosomes and RNA Metabolism" (Zelinka, J. and Balan, J., eds.), Vol. 2, pp. 227-236, Bratislava: Publishing House of the Slovak Academy of Sciences

CERNA, J., RICHLIK, I., KRAYEVSKY, A.A., GOTTIKH, B.P. (1973a) 2'(3')-O-(N-formylmethionyl)-adenosine-5'-phosphate. A new donor substrate in peptidyl transferase catalyzed reactions. FEBS Letters 37, 188-191

CERNA, J., RYCHLIK, I., KRAYEVSKY, A.A., GOTTIKH, B.P. (1973b) pA-Fmet, a new donor substrate for ribosomal peptidyl transferase. In "Ribosomes and RNA Metabolism" (Zelinka, J. and Balan, J., eds.), Vol. 1, pp. 285-295, Bratislava: Publishing House of the Slovak Academy of Sciences

CERNA, J., RYCHLIK, I., LICHTENTHALER, F.W. (1973) The effect of the aminoacyl-4-aminohexosyl-cytosine group of antibiotics on ribosomal peptidyl transferase. FEBS Letters 30, 147-150

CERNA, J.R., RYCHLIK, I., PULKRABEK, P. (1969) The effect of antibiotics on the coded binding of peptidyl-tRNA to the ribosome and on the transfer of the peptidyl residue to puromycin. Eur.J.Biochem. 9, 27-35

CHANG, F.N., FLAKS, J.G. (1972) The binding of dihydrostreptomycin to E. coli ribosomes: characteristics and equilibrium of the reaction. Ant.Agents Chemother. 2, 308-319

CHANG, T., NEVILLE, D.M. (1977) Artificial hybrid protein containing a toxic protein fragment and a cell membrane receptor-binding moiety in a disulfide conjugate. I. Synthesis of diphtheria toxin fragment A-S-S-human placental lactogen with methyl-5-bromovalerimidate. J.Biol.Chem. 252, 1505-1514

CHANG, F.N., SIDDHIKOL, C., WEISBLUM, B. (1969) Subunit localization studies of antibiotic inhibitors of protein synthesis. Biochim.Biophys.Acta 186, 396-398

CHANG, F.N., SIH, C.J., WEISBLUM, B. (1966) Lincomycin,an inhibitor of aminoacyl sRNA binding to ribosomes. Proc.Nat.Acad.Sci. 55, 431-438

CHANG, F.N., WEISBLUM, B. (1967) The specificity of lincomycin binding to ribosomes. Biochemistry 6, 836-843

CHAPEVILLE, F., GRUNBERG-MANAGO, M. (eds.) (1975) Organization and Expression of the Viral Genome. Molecular Interactions in Genetic Translation. Amsterdam-New York: North Holland-American Elsevier

CHEUNG, C.P., STEWART, M.L., GUPTA, N.K. (1973) Protein synthesis in rabbit reticulocytes: evidence for the synthesis of initial dipeptides in the presence of pactamycin. Biochem.Biophys.Res. Commun. 54, 1092-1101

CHINALI, G., WOLF, H., PARMEGGIANI, A. (1977) Effect of kirromycin on elongation factor Tu. Location of the catalytic center for ribosome·elongation-factor-Tu GTPase activity on the elongation factor. Eur.J.Biochem. 75, 55-65

CLARK, J.M., CHANG, A.Y. (1965) Inhibitors of the transfer of amino acids from aminoacyl soluble ribonucleic acid to proteins. J.Biol.Chem. 240, 4734-4739

CLARK, J.M., GUNTHER, J.K. (1963) Gougerotin, a specific inhibitor of protein synthesis. Biochim.Biophys.Acta 76, 636-638

CLEGG, J.C.S. (1975) Sequential translation of capsid and membrane protein genes of alphaviruses. Nature 254, 454-455

CLEMENS, M.J. (1976) Functional relationships between a reticulocyte polypeptide-chain-initiation factor (IF-MP) and the translational inhibitor involved in regulation of protein synthesis by haemin. Eur.J.Biochem. 66, 413-422

CLEMENS, M.J., HENSHAW, E.C., RAHAMIMOFF, H., LONDON, I.M. (1974) Met-$tRNA_f^{Met}$ binding to 40S ribosomal subunits: a site for the regulation of initiation of protein synthesis by hemin. Proc.Nat.Acad.Sci. 71, 2946-2950

CLEMENS, M.J., PAIN, V.M., HENSHAW, E.C., LONDON, I.M. (1976) Characterization of a macromolecular inhibitor of polypeptide chain initiation from Ehrlich ascites tumor cells. Biochem.Biophys. Res.Commun. 72, 768-775

COCITO, C., KAJI, A. (1971) Virginiamycin M, a specific inhibitor of the acceptor site of ribosomes. Biochimie 53, 763-770

COCITO, C., VOORMA, H.O., BOSCH, L. (1974) Interference of virginiamycin M with the initiation and the elongation of peptide chains in cell-free systems. Biochim.Biophys.Acta 340, 285-298

COCHRAN, J.W., BYRNE, R.W. (1974) Isolation and properties of a ribosome-bound factor required for ppGpp and pppGpp synthesis in Escherichia coli. J.Biol.Chem. 249, 353-360

CODDINGTON, A., FLURI, R. (1977) Characterization of the ribosomal proteins from Schizosaccharomyces pombe by two-dimensional polyacrylamide gel electrophoresis. Mol.Gen.Genetics 158, 93-100

COHEN, L.B., GOLDBERG, I.H., HERNER, A.E. (1969) Inhibition by pactamycin of the initiation of protein synthesis. Effect on the 30S ribosomal subunit. Biochemistry 8, 1327-1335

COHEN, L.B., HERNER, A.E., GOLDBERG, I.H. (1969) Inhibition by pactamycin of the initiation of protein synthesis. Binding of N-acetylphenylalanyl transfer ribonucleic acid and polyuridylic acid to ribosomes. Biochemistry 8, 1312-1326

Cold Spring Harb.Symp.Quant.Biol. (1969) The Mechanism of Protein Synthesis, Vol. 34

COLLIER, R.J. (1967) Effect of diphtheria toxin on protein synthesis: inactivation of one of the transfer factors. J.Mol.Biol. 25,83-98.

COLLIER, R.J. (1975) Diphtheria toxin: mode of action and structure. Bacteriol.Rev. 39, 54-85

COLLIER, R.J., PAPPENHEIMER, A.M. (1964) Studies on the mode of action of diphtheria toxin. II. Effect of toxin on amino acid incorporation in cell-free systems. J.Exp .Med. 120, 1020-1039

COLOMBO, B., FELICETTI, L., BAGLIONI, C. (1966) Inhibition of protein synthesis in rabbit reticulocytes on antibiotics. I. Effects on polysomes. Biochim.Biophys.Acta 119, 109-119

COLOMBO, B., VESCO, C., BAGLIONI, C. (1968) The role of ribosomal subunits in protein synthesis in mammalian cells. Proc.Nat.Acad. Sci. 61, 651-658

CONDE, F.P., FERNANDEZ-PUENTES, C., MONTERO, M.T.V., VAZQUEZ, D. (1978) Protein toxins that catalytically inactivate ribosomes from eukaryotic microorganisms. Studies on the mode of action of alpha sarcin, mitogillin and restrictocin: response to alpha sarcin antibodies. FEMS Microbiology Letters (in press)

CONDIT, R.C. (1975) F factor-mediated inhibition of bacteriophage levels following T7 infection of male *Escherichia coli*. J.Mol. Biol. 98, 45-46

CONNAMACHER, R.H., MANDEL, H.G. (1965) Binding of tetracycline to the 30S ribosomes and to polyuridylic acid. Biochem.Biophys.Res.Commun. 20, 98-103

CONNAMACHER, R.H., MANDEL, H.G. (1968) Studies on the intracellular localization of tetracycline in bacteria. Biochim.Biophys. Acta 166, 475-486

CONTRERAS, A., BARBACID, M., VAZQUEZ, D. (1974) Binding to ribosomes and mode of action of chloramphenicol analogues. Biochim.Biophys. Acta 349, 376-388

CONTRERAS, A., CARRASCO, L. (1978) Selective inhibition of picornavirus-infected cells by inhibitors of protein synthesis. J. Virology (in press)

CONTRERAS, A., VAZQUEZ, D. (1977a) Cooperative and antagonistic interactions with peptidyl-tRNA and antibiotics with bacterial ribosomes. Eur.J.Biochem. 74, 539-547

CONTRERAS, A., VAZQUEZ, D. (1977b) Synergistic interaction of the streptogramins with the ribosome. Eur.J.Biochem. 74, 549-551

CONTRERAS, A., VAZQUEZ, D., CARRASCO, L. (1978) Inhibition by selected antibiotics, of protein synthesis in cells growing in tissue cultures. J.Antibiotics 6, 151-155

CONTROULIS, J., REBSTOCK, M.C., CROOKS, H.M. (1949) Chloramphenicol (chloromycetin). V. Synthesis. J.Amer.Chem.Soc. 71, 2463-2468.

COOPER, D., BANTHORPE, D.V., WILKIE, D. (1967) Modified ribosomes conferring resistance to cycloheximide in mutants of Saccharomyces cerevisiae. J.Mol.Biol. 26, 347-350

COOPER, D.J., JARET, R.S., REIMAN, H. (1971) Structure of sisomicin, a novel unsaturated aminoglycoside antibiotic from Micromonospora myoensis. Chem.Commun.-1971, 285-286

COOPERMAN, B.S., JAYNES, E.N., BRUNSWICK, D.J., LUDDY, M.A. (1975) Photoincorporation of puromycin and N-(ethyl-2-diazomalonyl)-puromycin into Escherichia coli ribosomes. Proc.Nat.Acad.Sci. 72, 2974-2978

CORCORAN, J.W., HAHN, F.E. (eds.) (1975) Antibiotics III. Mechanism of Action of Antimicrobial and Antitumor Agents. Berlin-Heidelberg-New York: Springer-Verlag

CORDELL-STEWART, B., TAYLOR, M.(1973) Effect of viral double-stranded RNA on protein synthesis in intact cells. J.Virol. 11, 232-237

COUSIN, M.A., LANDO, D., OJASOO, T., RAYNAUD, J.P. (1977) Stability of 70S ribosomes in relation to misreading and antibacterial activity of aminoglycosides. Biochimie 59, 59-63

COUTSOGEORGOPOULOS, C. (1967) Aminoacyl aminonucleotides inhibitors of protein synthesis. The effect of aminoacyl ribonucleic acid on the inhibition. Biochemistry 6, 1704-1711

COX, E.C., WHITE, J.R., FLAKS, J.G. (1964) Streptomycin action and the ribosome. Proc.Nat.Acad.Sci. 51, 703-704

CRON, M.J., FARDIG, O.B., JOHNSON, D.L., WHITEHEAD, D.F., HOOPER, I.R., LEMIEUX, R.U. (1958) Kanamycin. I. Characterization and acid hydrolysis studies. J.Amer.Chem.Soc. 80, 4115

CUNDLIFFE, E. (1969) Antibiotics and polyribosomes. II. Some effects of lincomycin, spiramycin, and streptogramin A in vivo. Biochemistry 8, 2063-2066

CUNDLIFFE, E. (1971) The mode of action of thiostrepton in vivo. Biochem.Biophys.Res.Commun. 44, 912-917

CUNDLIFFE, E. (1972) The mode of action of fusidic acid. Biochem. Biophys.Res.Commun. 46, 1794-1801

CUNDLIFFE, E. (1978) Mechanism of resistance to thiostrepton in the producing-organism Streptomyces azureus. Nature 272, 792-795

CUNDLIFFE, E., CANNON, M., DAVIES, J. (1974) Mechanism of inhibition of eukaryotic protein synthesis by trichothecene fungal toxins. Proc.Nat.Acad.Sci. 71, 30-34

CUNDLIFFE, E., DAVIES, J.E. (1977) Inhibition of initiation elongation and termination of eukaryotic protein synthesis by trichothecene fungal toxins. Ant.Agents Chemother. 11, 491-499

CUNDLIFFE, E., DIXON, P. (1975) Inhibition of ribosomal A site functions by sporangiomycin and micrococcin. Ant.Agents Chemother. 8, 1-4

CUNDLIFFE, E., McQUILLEN, K. (1967) Bacterial protein synthesis: the effect of antibiotics. J.Mol.Biol. 30, 137-146

CZERNILOFSKY, P., COLLATZ, E., GRESSNER, A.M., WOOL, I.G., KUCHLER, E. (1977) Identification of the tRNA-binding sites on rat liver ribosomes by affinity labeling. Mol.Gen.Genetics 153, 231-235

DABBS, E.R. (1977) A spectinomycin dependent mutant of Escherichia coli. Mol.Gen.Genetics 151, 261-267

DAHLBERG, A.E., DAHLBERG, J.E. (1975) Binding of ribosomal protein S1 of Escherichia coli to the 3' end of 16S rRNA. (RNA-protein interaction/initiation of protein synthesis/colicin E3/ aurintricarboxylic acid). Proc.Nat.Acad.Sci. 72, 2940-2944

DAHLBERG, A.E., LUND, E., KJELDGAARD, N.O., BOWMAN, C.M., NOMURA, M. (1973) Colicin E3 induced cleavage of 16S ribosomal ribonucleic acid; blocking effects of certain antibiotics. Biochemistry 12, 948-950

DALLAL, J.A., IRVIN, J.D. (1978) Enzymatic inactivation of eukaryotic ribosomes by the pokeweed antiviral protein. FEBS Letters 89, 257-259

DANIELS, P.J.L., KUGELMAN, M., MALLAMS, K., TKACH, R.W., VERNAY, H.F., WEINSTEIN, V., YEHASKEL, A. (1971) Mass-spectral studies on aminocyclitol antibiotics. Chem.Commun. 1629-1631

DARNBROUGH, C., HUNT, T., JACKSON, R.J. (1972) A complex between Met-$tRNA_F$ and native 40S subunits in reticulocyte lysates and its disappearance during incubation with double-stranded RNA. Biochem.Biophys.Res.Commun. 48, 1556-1564

DAS, H.K., GOLDSTEIN, A., KANNER, L.C. (1966) Inhibition by chloramphenicol of the growth of nascent protein chains in Escherichia coli. Mol.Pharmacol. 2, 158-162

DATTA, A., de HARO, C., SIERRA, J.M., OCHOA, S. (1977a) Role of 3': 5'-cyclic-AMP-dependent protein kinase in regulation of protein synthesis in reticulocyte lysates. Proc.Nat.Acad.Sci. 74, 1463-1467

DATTA, A., de HARO, C., SIERRA, J.M., OCHOA, S. (1977b) Mechanism of translational control by hemin in reticulocyte lysates. Proc.Nat. Acad.Sci. 74, 3326-3329

DAVIES, J.E. (1964) Studies on the ribosomes of streptomycin sensitive and resistant strains of Escherichia coli. Proc.Nat.Acad.Sci. 51, 659-664

DAVIES, J. (1966) Streptomycin and the genetic code. Cold Spring Harb.Symp.Quant.Biol. 31, 657-664

DAVIES, J. (1970) Structure-activity relationship among the aminoglycoside antibiotics: comparison of the neomycins and hybrimycins. Biochim.Biophys.Acta 222, 674-676

DAVIES, J., ANDERSON, P., DAVIS, B.D. (1965) Inhibition of protein synthesis by spectinomycin. Science 149, 1096-1098

DAVIES, J., ANDERSON, P., DAVIS, B.D. (1967) Effect of spectinomycin on polypeptide synthesis in extracts by Escherichia coli. J.Mol. Biol. 29, 203-215

DAVIES, J., DAVIS, B.D. (1968) Misreading of ribonucleic acid code words induced by aminoglycoside antibiotics. J.Biol.Chem. 243, 3312-3316

DAVIES, J.E., GILBERT, W., GORINI, L. (1964) Streptomycin, suppression and the code. Proc.Nat.Acad.Sci. 51, 883-890

DAVIES, J., GORINI, L., DAVIS, B.D. (1965) Misreading of RNA codewords induced by aminoglycoside antibiotics. Mol.Pharmacol. 1, 93-106

DAY, L.E. (1966) Tetracycline inhibition of cell-free protein synthesis. II. Effect of the binding of tetracycline to the components of the system. J.Bacteriol. 92, 197-203

DE BOER, C., DIETZ, A.,WILKINS, J.R., LEWIS, C.N., SAVAGE, G.M. (1955) Celesticetin a new crystalline antibiotic. I. Biologic studies of celecticetin. Antibiotics Annual 1954-1955, 832-836

DEKIO, S., TANAKA, R., OSAWA, S., TANAKA, K., TAMAKI, M. (1970) Genetic studies of the ribosomal proteins in Escherichia coli. IV. Puttern of the alteration of ribosomal protein components in mutants resistant to spectinomycin or erythromycin in different strains of Escherichia coli. Mol.Gen.Genetics 107, 39-49

DELIHAS, N., TOPOL, E., LARRINUA, I. (1975) Changes in accessibility to chemical modification of Escherichia coli ribosomes induced by streptomycin. FEBS Letters 53, 170-175

DELUNGE, R.J., DRAZIN, R.E., COLLIER, R.J. (1976) Aminoacid sequence of fragment A, an enzymically active fragment from diphtheria toxin. Proc.Nat.Acad.Sci. 73, 69-72

DIXON, P.D., BEVEN, J.E., CUNDLIFFE, E. (1975) Properties of the ribosomes of antibiotic producers: effects of thiostrepton and micrococcin on the organisms which produce them. Ant.Agents Chemother. 7, 850-855

DONALDSON, G.R., ATKINSON, M.R., MURRAY, A.W. (1968) Inhibition of protein synthesis in Ehrlich ascites-tumour cells by the phenanthrene alkaloids tylophorine, tylocrebrine and cryptopleurine. Biochem.Biophys.Res.Commun. 31, 104-109

DUBNAU, D., GOLDTHWAITE, C., SMITH, I., MARMUR, J. (1967) Genetic mapping in Bacillus subtilis. J.Biol.Chem. 27, 163-185

DUGGAR, B.M. (1948) Aureomycin, a product of the continuing search for new antibiotics. Ann.New York Acad.Sci. 51, 177-181

DUNITZ, J.D. (1952) The crystal structure of chloramphenicol and bromanphenicol. J.Amer.Chem.Soc. 74, 995-999

DUQUETTE, P.H., RITTER, C.L., VINCE, R. (1974) Puromycin analogs. Ribosomal binding and peptidyl transferase substrate activity of a carbocyclic analog of 8-azapuromycin. Biochemistry 13, 4855-4859

DUTCHER, J.D., DONIN, M.N. (1952) The identity of neomycin A, neamine and methanolysis product of neomycin B and C. J.Amer. Chem.Soc. 74, 3420-3422

ECKERMANN, D.J., SYMONDS, R.H. (1978) Sequence at the site of attachment of an affinity-label derivative of puromycin on 23-S ribosomal RNA of Escherichia coli ribosomes. Eur.J.Biochem. 82, 225-234

ECKSTEIN, F., KETTLER, M., PARMEGGIANI, A. (1971) Guanylylimido-diphosphate and its interaction with amino acid polymerization factor. Biochem.Biophys.Res.Commun. 45, 1151-1158

EGBERTS, E., HACKETT, P.B., TRAUB, P. (1977) Alteration of the intracellular energetic and ionic conditions by mengovirus infection of Ehrlich ascites tumor cells and its influence on protein synthesis in the midphase of infection. J.Virol. 22, 591-597

EGGER, H., REINSHAGEN, H. (1976a) New pleuromutilin derivatives with enhanced antimicrobial activity. I. Synthesis. J.Antibiotics 29, 915-922

EGGER, H., REINSHAGEN, H. (1976b) New pleuromutilin derivatives with enhanced antimicrobial activity. II. Structure-activity correlations. J.Antibiotics 29, 923-927

EHRENFELD, E., HUNT, T. (1971) Double-stranded poliovirus RNA inhibits initiation of protein synthesis by reticulocyte lysates. Proc.Nat.Acad.Sci. 68, 1075-1078

EHRLICH, P. (1957) The Collected Papers of Paul Ehrlich, Vol. 2, pp. 21-44. Pergamon Press: London

EHRLICH, J., BARTZ, Q.R., SMITH, R.M., JOSLYN, D.A., BURKHOLDER, P.A. (1947) Chloromycetin, a new antibiotic from a soil actinomycete. Science 106, 417

EIL, C., WOOL, I.G. (1973) Function of phosphorylated ribosomes. The activity of ribosomal subunits phosphorylated in vitro by protein kinase. J.Biol.Chem. 248, 5130-5136

EISENHUTH, W., STAMM, O.A., SCHMID, H. (1964) Uber die struktur des chartewsins II. Helv.Chim.Acta 47, 1475-1484

ENDO, T., YONEHARA, H. (1978) Identity of multhiomycin and nosiheptide 31, 623-625

ENGLAND, J.M., HOWETT, M.K., TAN, K.B. (1975) Effect of hypertonic conditions on protein synthesis in cells productively infected with Simian virus 40. J.Virol. 16, 1101-1107

ENNIS, H.L. (1965a) Inhibition of polypeptide synthesis by polypeptide antibiotics. I. Inhibition in intact bacteria. J.Bacteriol. 90, 1102-1108

ENNIS, H.L. (1965b) Inhibition of protein synthesis by polypeptide antibiotics. II. In vitro protein synthesis. J.Bacteriol. 90, 1109-1119

ENNIS, H.L. (1966) Inhibition of protein synthesis by polypeptide antibiotics. III. Ribosomal site of inhibition. Mol.Pharmacol. 2, 444-452

ENNIS, H.L. (1970) In "Progress in Antimicrobial and Anticancer Chemotherapy", Vol. II, pp. 489-498, Tokyo: University of Tokyo Press

ENNIS, H.L. (1971) Interaction of vernamycin A with Escherichia coli ribosomes. Biochemistry 10, 1265-1270

ENNIS, H.L. (1972) Polysome metabolism in Escherichia coli: effect of antibiotics on polysome stability. Ant.Agents Chemother. 1, 197-203

ENNIS, H.L. (1974) Binding of the antibiotic vernamycin in Bα to Escherichia coli ribosomes. Arch.Biochem.Biophys. 160, 394-401

ENNIS, H.L., DUFFY, K.E. (1972) Vernamycin A inhibits the non-enzymic binding of Fmet-tRNA to ribosomes. Biochim.Biophys.Acta 281, 93-102

ENNIS, H.L., LUBIN, M. (1964) Cycloheximide: aspects of inhibition of protein synthesis in mammalian cells. Science 146, 1474-1476

ERDÖS, T., ULLMANN, A. (1959) Effect of streptomycin on the incorporation of amino acids labelled with carbon-14 into ribonucleic acid and protein in a cell-free system of a mycobacterium. Nature, 183, 618-619

EVERSE, J., GARDNER, D.A., KAPLAN, N.O., GALASINSKI, W., MOLDAVE, K. (1970) The formation of a ternary complex between diphtheria toxin, aminoacyl-transferase II, and diphosphopyridine nucleotide. J.Biol.Chem. 245, 899-905

FAHNESTOCK, S., NEUMANN, H., SHASHOUA, V., RICH, A. (1970) Ribosome-catalyzed ester formation. Biochemistry 9, 2477-2483

FALCOFF, E., FALCOFF, R., LEBLEU, B., REVEL, M. (1973) Correlation between the antiviral effect of interferon treatment and the inhibition of in vitro mRNA translation in noninfected L cells. J.Virol. 12, 421-430

FARKAS, G., ANTONI, F., STAUB, M., PIFFKO, P. (1974) Emetine and macromolecular biosynthesis in mammalian cells. Acta Biochim. Biophys.Sci.Hung. 9, 63-72

FARRELL, P.J., BALKOW, K., HUNT, T., JACKSON, R.J., TRASCHEL, N. (1977) Phosphorylation of initiation factor eIF-2 and the control of reticulocyte protein. Cell 11, 187-200

FELICETTI, L., COLOMBO, B., BAGLIONI, C. (1966) Inhibition of protein synthesis in reticulocytes by antibiotics. II. The site of action of cycloheximide, streptovitacin A and pactamycin. Biochim.Biophys.Acta 119, 120-129

FERNANDEZ-MUÑOZ, R., MONRO, R.E., TORRES-PINEDO, R., VAZQUEZ, D. (1971) Substrate- and antibiotic-binding sites at the peptidyl-transferase centre of E. coli ribosomes. Studies on the chloramphenicol, lincomycin and erythromycin sites. Eur.J. Biochem. 23, 185-193

FERNANDEZ-MUÑOZ, R., VAZQUEZ, D. (1973a) Binding of puromycin to Escherichia coli ribosomes. Effects of puromycin analogues and peptide bond formation inhibitors. Mol.Biol.Reports 1, 27-32

FERNANDEZ-MUÑOZ, R., VAZQUEZ, D. (1973b) Kinetic studies of peptide bond formation. Effect of chloramphenicol. Mol.Biol.Reports 1, 75-79

FERNANDEZ-MUÑOZ, R., VAZQUEZ, D. (1973c) Quantitative binding of ^{14}C-erythromycin to E. coli ribosomes. J.Antibiotics 26, 107-108

FERNANDEZ-PUENTES, C., BENSON, S., OLSNES, S., PIHL, A. (1976) Protective effect of elongation factor 2 on the inactivation of ribosomes by the toxic lectins abrin and ricin. Eur.J.Biochem. 64, 437-443

FERNANDEZ-PUENTES, C., CARRASCO, L., VAZQUEZ, D. (1976) Site of action of ricin on the ribosome. Biochemistry 15, 4364-4369

FERNANDEZ-PUENTES, C., VAZQUEZ, D. (1977) Effects of some proteins that inactivate the eukaryotic ribosome. FEBS Letters 78, 143-146

FEY, G., REISS, M., KERSTEN, H. (1973) Interaction of tetracyclines with ribosomal subunits from Escherichia coli. A fluorometric investigation. Biochemistry 12, 1160-1164

FINLAY, A.C., HOBBY, G.L., P'AN, S.Y., REGNA, P.P., ROUTIEN, J.B., SEELEY, D.B., SHULL, G.M., SOBIN, B.A., SOLOMONS, I.A., WINSON, J.W., KANE, J.H. (1950) Science 11, 85

FINLAY, A.C., HOBBY, G.L., HOCHSTEIN, F., LEES, T.M., LENERT, T.F., MEANS, J.A., P'AN, S.Y., REGNA, P.P., ROUTIEN, J.B., SOBIN, B.A., TATE, K.B., KANE, J.H. (1951) Viomycin, a new antibiotic active against mycobacteria. Amer.Rev.Tuberc. 63, 1-6

FIRKIN, F.C., LINNANE, A.W. (1969) Phylogenetic differences in the sensitivity of mitochondrial protein synthesising systems to antibiotics. FEBS Letters 2, 330-332

FISCHER, E., WOLF, H.,HANTKE, K., PARMEGGIANI, A. (1977) Elongation factor Tu resistant to kirromycin in an *Escherichia coli* mutant altered in both tuf genes. Proc.Nat.Acad.Sci. 74, 4341-4345

FLAKS, J.G., COX, E.C., WITTING, M.L., WHITE, J.R. (1962) Polypeptide synthesis with ribosomes from streptomycin-resistant and dependent *E. coli*. Biochem.Biophys.Res.Commun. 7, 390-393

FODSTAD, O., OLSNES, O. (1977). Studies on the accessability of ribosomes to inactivation by the toxic lectins abrin and ricin. Eur.J.Biochem. 74, 209-215

FORTNAGEL, P., BERGMAN, R. (1973) Alteration of the ribosomal fraction of *Bacillus subtilis* during sporulation. Biochim. Biophys.Acta 299, 136-141

FORTNAGEL, P., FREESE, E.B. (1977) Morphological stages of *Bacillus subtilis* sporulation and resistance to fusidic acid. J.Gen. Microbiol. 101, 299-306

FOX, J.J., WATANABE, S. (1966) Nucleosides XXXII. On the structure of blasticidin S, a nucleoside antibiotic. Tetrahedron Letters 897-904

FRANKLIN, T.J. (1963) The inhibition of incorporation of leucine into protein of cell-free systems from rat liver and *Escherichia coli* by chlortetracycline. Biochem.J. 87, 449-453

FREEMAN, K.B., PATEL, H., HALDAR, D. (1977) Inhibition of deoxiribonucleic acid synthesis in Ehrlich ascites cells by chloramphenicol. Mol.Pharmacol. 13, 504-511

FRESNO, M., CARRASCO, L., VAZQUEZ, D. (1976) Initiation of the polypeptide chain by reticulocyte cell-free systems. Survey of different inhibitors of translation. Eur.J.Biochem. 68, 355-364

FRESNO, M., GONZALEZ, A., VAZQUEZ, D., JIMENEZ, A. (1978) Bruceantin-a novel inhibitor of peptide bond formation. Biochim.Biophys. Acta 518, 104-112

FRESNO, M., JIMENEZ, A., VAZQUEZ, D. (1977) Inhibition of translation in eukaryotic systems by harringtonine. Eur.J. Biochem. 72, 323-330

FRESNO, M., VAZQUEZ, D. (1978) Initiation of translation with native 40S ribosomal subunits. Eur.J.Biochem. 83, 169-178

FRIEDMAN, H., LU, P., RICH, A. (1969) An in vivo block in the initiation of protein synthesis. Cold Spring Harb.Symp.Quant. Biol. 34, 255-260

FRIEDMAN, R.M., METZ, D.H., ESTEBAN, R.M., TOVELL, D.R., BALL, L.A., KERR, I.M. (1972) Mechanism of interferon action: inhibition of viral messenger ribonucleic acid translation in L-cell extracts. J.Virol. 10, 1184-1198

FRIEDMAN, S.M., WEINSTEIN, I.B. (1964) Lack of fidelity in the translation of synthetic polyribonucleotides. Proc.Nat.Acad.Sci. 52, 988-996

FUGANTI, C. (1975) In "The Alkaloids" (Manske, R.H.F., ed .), Vol. 15, Chapter 3, pp. 83-164, New York: Academic Press

FUJIMOTO, H., KINOSHITA, T., SUZUKI, H., UMEZAWA, H. (1970) Studies on the mode of action of althiomycin. J.Antibiotics 23, 271-275

FUNATSU, M. (1972) In "Proteins, Structure and Function" (Funatsu, M., Hiromi, K., Imahori, K., Murachi, T., Narita, K., eds.), pp. 103-139, New York: John Wiley & Sons

FUNATSU, G., NIERHAUS, K., WITTMANN-LIEBOLD, B. (1972) Ribosomal proteins. XXII. Studies on the altered protein S5 from a spectinomycin resistant mutant of Escherichia coli. J.Mol.Biol. 64, 201-209

FUNATSU, G., NIERHAUS, K., WITTMANN, H.G. (1972) Ribosomal proteins XXXVII. Determination of allele types and amino acid exchanges in protein S12 of three streptomycin-resistant mutants of Escherichia coli. Biochim.Biophys.Acta 287, 282-291

FUNATSU, G., WITTMANN, H.G. (1972) Ribosomal proteins. XXXIII. Location of amino acid replacements in protein S12 of three streptomycin-resistant mutants of Escherichia coli. Biochim. Biophys.Acta 287,282-291

FURUSAWA, E., CUTTING, W. (1970) The higher plants with antiviral and anti-lethal activity on virus infections in mice. Ann.New York Acad.Sci. 173, 668-679

FURUSAWA, E., FURUSAWA, S., LEE, J.Y.B., PATANAVANICH, S. (1976) Therapeutic activity of pretazzetine, a Narcissus alkaloid on rauscher leukemia: comparison with tazettine and streptonigrin. Proc.Soc.Exp.Biol.Med. 152, 186-191

GALE, E.F. (1963) Mechanism of antibiotic action. Pharmacol.Rev. 15, 481-530

GALE, E.F., CUNDLIFFE, E., REYNOLDS, P.E., RICHMOND, M.H., WARING, M.J. (1972) In "The Molecular Basis of Antibiotic Action", pp. 278-379, London: John Wiley & Sons

GALE, E.F., FOLKES, J.P. (1953) The assimilation of aminoacids by bacteria. 15. Actions of antibiotics on nucleic acid and protein synthesis in Staphylococcus aureus. Biochem.J. 53, 493-498

GALE, E.F., PAINE, T.F. (1950a) Glutamic acid accumulation and protein synthesis in Staphylococcus aureus. Biochem.J. 47, XXVI

GALE, E.F., PAINE, T.F. (1950b) The assimilation of aminoacids by bacteria. 12. The action of inhibitors and antibiotics on the accumulation of free glutamic acid and the formation of combined glutamate in Staphylococcus aureus. Biochem.J. 48, 298-301

GARCIA-PATRONE, M., PERAZZOLO, C.A., BARALLE, F., GONZALEZ, N.S., ALGRANATI, I.D. (1971) Studies on dissociation factor of bacterial ribosomes: effect of antibiotics. Biochim.Biophys. Acta 246, 291-299

GAREN, A., SIDDIKI, O. (1962) Suppression of mutations in the alkaline phosphatase structural cistron of E. coli. Proc.Nat.Acad. Sci. 48, 1121-1127

GARVIN, R.T., BISWAS, D.K., GORINI, L. (1974) The effects of streptomycin or dihydrostreptomycin binding to 16S RNA or to 30S ribosomal subunits. Proc.Nat.Acad.Sci. 71, 3814-3818

GASPERI-CAMPANI, A., BARBIERI, L., LORENZONI, E., MONTANARO, L., SPERTI, S., BONETTI, E., STIRPE, F. (1978) Purification, toxicity and inhibition of protein synthesis in vitro. Biochem. J. (in press)

GILL, D.M., PAPPENHEIMER, A.M., BASEMAN, J.B. (1969) Studies on transferase II using diphtheria toxin. Cold Spring Harb. Symp.Quant.Biol. 34, 595-602

GIRBES, T., VAZQUEZ, D., MODOLELL, J. (1976) Polypeptide-chain elongation promoted by guanyl-5'-yl imidodiphosphate. Eur.J. Biochem. 67, 257-265

GIRSHOVICH, A.S., BOCHKAREVA, E.S., OVCHINNIKOV, Y.A. (1976) Identification of components of the streptomycin-binding center of E. coli MRE600 ribosomes by photo-affinity labelling. Mol. Gen.Genetics 144, 205-212

GLASBY, J.S. (1976) Encyclopedia of Antibiotics. John Wiley & Sons: London-New York-Sydney-Toronto

GODCHAUX, W. III., ADAMSON, S.D., HERBERT, E. (1967) Effects of cycloheximide on polyribosome function in reticulocytes. J.Mol. Biol. 27, 57-72

GODFREDSEN, W.O., DAEHNE, W., VANGEDAL, S. (1965) The stereo chemistry of fusidic acid. Tetrahedron 21, 3505-3530

GODFREDSEN, W.O., HANNSEN, S., LORCK, H., ROHOLT, K., TYBRING, L. (1962) Fusidic acid: a new antibiotic. Nature 193, 987

GODFREDSEN, W.O., VANGEDAL, S. (1964) Trichodermin a new antibiotic, related to trichothecin. Proc.Chem.Soc. 188-199

GOFFIC, F.le, CAPMAU, M.L., BONNET, D., CERCEAU, C., SOUSSY, C. DUBLANCHET, A., DUVAL, J. (1977) Plasmid mediated pristinamycin resistance PAC IIA: a new enzyme which modified pristinamycin IIA. J.Antibiotics 30, 665-669

GOLDBERG, I.H., MITSUGI, K. (1966) Sparsomycin, an inhibitor of aminoacyl transfer to polypeptide. Biochem.Biophys.Res.Commun. 23, 453-459

GOLDBERG, I.H., MITSUGI, K. (1967a) Inhibition by sparsomycin and other antibiotics of the puromycin-induced release of polypeptide from ribosomes. Biochemistry 6, 383-390

GOLDBERG, I.H., MITSUGI, K. (1967b) Sparsomycin inhibition of polypeptide synthesis promoted by synthetic and natural polynucleotides. Biochemistry 6, 372-382

GOLDBERG, I.H., STEWART, M.L., AYUSO, M., KAPPEN, L.S. (1973) On the mechanisms of inhibition of polypeptide synthesis by the antibiotics sparsomycin and pactamycin. Federation Proc. 32, 1688-1697

GOLDBERG, R., RINGER, D., CHLADEK, S. (1977) Inhibition of chloramphenicol binding to Escherichia coli 70S ribosomes by 2'(3')-O-aminoacyl-dinucleoside phosphates derived from the aminoacyl-tRNA acceptor terminus. Eur.J.Biochem. 81, 373-378

GOLDEMBERG, S.H., ALGRANATI, I.D. (1977) Polyamines and protein synthesis: studies in various polyamine-requiring mutants of Escherichia coli. Mol.Cell.Biochem. 16, 71-77

GOLDSTEIN, J., WALDMAN, A.A., MARX, G. (1975) Heparin as an inhibitor of mammalian protein synthesis. Adv.Exp.Med.Biol. 52, 289-297

GOLDTHWAITE, C., SMITH, I. (1972) Physiological characterization of antibiotic resistant mutants of Bacillus subtilis. Mol.Gen. Genetics 114, 190-204

GONZALEZ, A., JIMENEZ, A., VAZQUEZ, D., DAVIES, J.E., SCHINDLER,D. (1978) Studies on the mode of action of hygromycin B: an inhibitor of translocation by eukaryotes. Biochim.Biophys.Acta (in press)

GONZALEZ, A., VAZQUEZ, D., JIMENEZ, A. (1978) Anthelmycin (hikizimycin), an inhibitor of translation by bacterial and eukaryotic systems. Biochim.Biophys.Acta (in press)

GONZALEZ-CADAVID, N.F., HERRERA QUIJADA, F. (1974) Inhibition of translation in liver polyribosomes by a new substituted thiospseudourea with antitumour action. Biochem.J. 138, 129-141

GOOCH, J., HAWTREY, A.O. (1975) Synthesis of thiol-containing analogues of puromycin and a study of their interaction with N-acetylphenylalanyl-transfer ribonucleic acid on ribosomes to form thioesters. Biochem.J. 149, 209-220

GOODWIN, F., SHAFRITZ, D., WEISSBACH, H. (1969) In vitro polypeptide synthesis in brain. Arch.Biochem.Biophys. 130, 183-190

GOOR, R.S., PAPPENHEIMER, A.M. (1967) Studies on the mode of action of diphtheria toxin. III. Site of toxin action in cell-free extracts. J.Exp.Med. 126, 899-912

GORDON, J. (1969) Hydrolysis of guanosine 5'-triphosphate associated with binding of aminoacyl transfer ribonucleic acid to ribosomes. J.Biol.Chem. 244, 5680-5686

GORINI, L., KATAJA, E. (1964) Phenotypic repair by streptomycin of defective genotypes in E. coli. Proc.Nat.Acad.Sci. 51, 487-493

GORINI, L., KATAJA, E. (1965) Supression activated by streptomycin and related antibiotics in drug-sensitive strains. Biochim. Biophys.Res.Commun. 18, 656-663

GOTTESMAN, M.E. (1967) Reaction of ribosome-bound peptidyl transfer ribonucleic acid with puromycin. J.Biol.Chem. 242, 5564-5571

GOTTLIEB, D., SHAW, P.D. (eds.) (1967a) Antibiotics, Vol. I. Berlin-Heidelberg-New York: Springer-Verlag

GOTTLIEB, D., SHAW, P.D. (eds.) (1967b) Antibiotics, Vol. II. Biosynthesis. Berlin-Heidelberg-New York: Springer-Verlag

GOTTLIEB, C., SKINNER, S.A.M., KORNFELD, S. (1974) Isolation of a clone of chinese hamster ovary cells deficient in plant lectin-binding sites. Proc.Nat.Acad.Sci. 71, 1078-1082

GRAAF, F.K.de., NIEKUS, H.G.D., KLOOTWIJK, J. (1973) Inactivation of bacterial ribosomes in vivo and in vitro by cloacin DF13. FEBS Letters 35, 161-165

GRAAF, F.K.de., PLANTA, R.J., STOUTHAMER, A.H. (1971) Effect of a bacteriocin produced by Enterobacter cloacae on protein biosynthesis. Biochim.Biophys.Acta 240, 122-136

GRANDI, M., HELMA, A., KUNTZEL, H. (1971) Fusidic acid resistance of mitocondrial G factor from Neurospora crassa. Biochem.Biophys. Res.Commun. 44, 864-871

GRANT, P., SANCHEZ, L., JIMENEZ, A. (1974) Cryptopleurine resistance: genetic locus for a 40S ribosomal component in Saccharomyces cerevisiae. J.Bacteriol. 120, 1308-1314

GRANT, P.G., SCHINDLER, D., DAVIES, J.E. (1976) Mapping of trichodermin resistance in Saccharomyces cerevisiae: a genetic locus for a component of the 60S ribosomal subunit. Genetics 83, 667-673

GRAVELA, E., POLI, G. (1977) Phalloidin poisoning of the isolated hepotocytes: inhibition of protein synthesis. Experientia 33, 603-604

GRECO, M., MONTANARO, L., NOVELLO, F., SACCONE, C., SPERTI, S., STIRPE, F. (1974) Inhibition of protein synthesis by ricin: experiments with rat liver mitochondria and nucleic and with ribosomes from Escherichia coli. Biochem.J. 142, 695-697

GREENWELL, P., HARRIS, R.J., SYMONDS, R.H. (1974) Affinity labelling of 23-S ribosomal RNA in the active centre of Escherichia coli peptidyl transferase. Eur.J.Biochem. 49, 539-554

GREGG, R.E., HEINTZ, R.L. (1972) The inhibition of eukaryotic aminoacyl transferase I by chartreusin. Arch.Biochem.Biophys. 152, 451-456

GRESSNER, A.M., WOOL, I.G. (1974) The stimulation of the phosphorylation of ribosomal protein S6 by cycloheximide and puromycin. Biochem. Biophys.Res.Commun. 60, 1482-1489

GRESSNER, A.M., WOOL, I.G. (1976) Effect of experimental diabetes and insulin on phosphorylation of rat liver ribosomal protein S6. Nature 259, 148-150

GRILL, L.K., SUN, J.D., KANDEL, J. (1976) Effect of double stranded RNA on protein synthesis in an in vitro wheat germ embryo system. Biochem.Biophys.Res.Commun. 73, 149-156

GROLLMAN, A.P. (1966) Structural basis for inhibition of protein synthesis by emetine and cycloheximide based on an analogy between ipecac alkaloids and glutarimide antibiotics. Proc.Nat. Acad.Sci. 56, 1867-1874

GROLLMAN, A.P. (1967a) Mode of action of anisomycin. J.Biol.Chem. 242, 3226-3233

GROLLMAN, A.P. (1967b) Structural basis for the inhibition of protein biosynthesis: mode of action of tubulosine. Science 157, 84-85

GROLLMAN, A.P. (1968) Inhibitors of protein synthesis. V. Effects of emetine on protein and nucleic acid biosynthesis in HeLa cells. J.Biol.Chem. 243, 4089-4094

GROLLMAN, A.P., STEWART, M.L. (1968) Inhibition of the attachment of messenger ribonucleic acid to ribosomes. Proc.Nat.Acad.Sci. 61, 719-725

GROLLMAN, A.P., GRUNFELD, C., BREWER, C.F., MARCUS, D.M. (1974) Molecular pharmacology of plant lecting: studies on ricin and concanavalin A (NSC-143504). Cancer Chemother.Rep. 58, 491-508

GROSS, M., RABINOVITZ, M. (1972) Controol of globin synthesis by hemin: factors influencing formation of an inhibitor of globin chain initiation in reticulocyte lysates. Biochim.Biophys.Acta 287, 340-352

GUARINO, A.J., IBERSHOF, M.L., SWAIN, R. (1963) Inhibition of protein synthesis in cell-free systems by homocitrullylamino adenosine. Biochim.Biophys.Acta 72, 62-68

GRUNBERG-MANAGO, M., DONDON, J., GRAFFE, M. (1972) Inhibition by thiostrepton of the IF 2-dependent ribosomal GTPase. FEBS Letters 22, 217-221

GUPTA, S.L., SOPORI, M.L., LENGYEL, P. (1973) Inhibition of protein synthesis directed by added viral and cellular messenger RNAs in extracts of interferon-treated Ehrlich ascites tumor cells. Location and dominance of the inhibitor(s). Biochim.Biophys.Res. Commun. 54, 777-783

GUPTA, R.S., SIMINOVITCH, L. (1977a) Mutants of CHO cells resistant to the protein synthesis inhibitors, cryptopleurine and tylocrebrine: genetic and biochemical evidence for common site of action of emetine, cryptopleurine, tylocrebrine and tubulosine. Biochemistry 16, 3209-3214

GUPTA, R.S., SIMINOVITCH, L. (1977b) The molecular basis of emetine resistance in chinese hamster ovary cells: alteration in the 40S ribosomal subunit.Cell 10, 61-66

HAENNI, A.L., LUCAS-LENARD, J. (1968) Stepwise synthesis of a tripeptide. Proc.Nat.Acad.Sci. 61, 1363-1369

HAMADA, M., HASHIMOTO, T., TAKAHASHI, T., YOKOYAMA, S., MIYAKE, M., TAKEUCHI, T., OKAMI, Y., UMEZAWA, H. (1965) Antimicrobial activity of kasugamycin. J.Antibiotics 18, 104-106

HAMILL, R.L., HOEHN, M.M. (1964) Anthelmycin, a new antibiotic with anthelmintinc properties. J.Antibiotics 17, 100-103

HARBON, S., CHAPEVILLE, F. (1970) Inhibition of protein synthesis in the reticulocytes by aminoacyl-adenosine. Eur.J.Biochem. 13, 375-383

HARRIS, R.J., HANLON, J.E., SYMONDS, R.H. (1971) Peptide bond formation on the ribosome. Structural requirements for inhibition of protein synthesis and of release of peptides from peptidyl-tRNA on bacterial and mammalian ribosomes by aminoacyl-tRNA and nucleotidyl analogues. Biochim.Biophys.Acta 240, 244-262

HASELKORN, R., ROTHMAN-DENES, L.B. (1973) Protein synthesis. Ann.Rev. Biochem. 42, 397-437

HASENBANK, R., GUTHRIE, C., STOFFLER, G., WITTMANN, H.G., ROSEN, L., APIRION, D. (1973) Electrophoretic and immunological studies on ribosomal proteins of 100 Escherichia coli revertants from streptomycin dependence. Mol.Gen.Genetics 127, 1-18

HASKELL, T.H., RYDER, A., FROHARDT, R.P., FUSARI, S.A., JAKUBOWSKI, Z.L., BARTZ, Q.R. (1958) The isolation and characterization of three crystalline antibiotics from Streptomyces plicatus. J.Amer. Chem.Soc. 80, 743-747

HASLAM, J.M., DAVEY, P.J., LINNANE, W. (1968) Differentiation in vitro by phenanthrene alkaloids of yeast mitochondrial protein synthesis from ribosomal systems of both yeast and bacteria. Biochem.Biophys.Res.Commun. 33, 368-373

HAUPT, I., FRICKE, H., CERNA, J., RYCHLIK, I. (1976) Effect of the leucomycin like macrolide antibiotic turimycin on ribosomal peptidyltransferase from Escherichia coli. J. Antibiotics 29, 1314-1319

HAUPT, I., FRICKE, H., CERNA, J., RYCHLIK, I. (1978) Effect of demycarosylturimycin on ribosomal peptidyltransferase from Escherichia coli. J. Antibiotics 31, 610-615

HEDBLOM, M.L., CAWLEY, D.B., HOUSTON, L.L. (1976) The specific binding of ricin and its polypeptide chains to rat liver ribosomes and ribosomal subunits. Arch.Biochem.Biophys. 177, 46-55

HEIBERG, R., OLSNES, S., PIHL, A. (1971) On the mechanism of the inhibiting effect of aurintricarboxylic acid on the binding of ribosomes to ribonucleic acids. FEBS Letters 18, 169-171

HELGELAND, K. (1976) Effect of fluoride on protein and collagen biosynthesis in rabbit dental pulp in vitro. Scand.J.Dent.Res. 84, 276-285

HELSER, T.L., DAVIES, J.E., DAHLBERG, J.E. (1971) Change in methylation of 16S ribosomal RNA associated with mutation to kasugamycin resistance in Escherichia coli. Nature New Biol. 233, 12-14

HELSER, T.L., DAVIES, J.E., DAHLBERG, J.E. (1972) Mechanism of kasugamycin resistance in Escherichia coli. Nature New Biol. 235, 6-9

HERNER, A.E., GOLDBERG, I.H., COHEN, L.B. (1969) Stabilization of N-acetylphenylalanyl transfer ribonucleic acid binding to ribosomes by sparsomycin. Biochemistry 8, 1335-1344

HERR, E.B.Jr. (1962) Chemical and biological properties of capreomycin and other peptide antibiotics. Ant.Agents Chemother. 201-212

HERRLICH, P., SCHWEIGER, M. (1976) Nitrofurans, a group of synthetic antibiotics, with a new mode of action: discrimination of specific messenger RNA classes. Proc.Nat.Acad.Sci. 73, 3386-3390

HERSHEY, J.W.B., MONRO, R.E. (1966) A competitive inhibitor of the GTP reaction in protein synthesis. J.Mol.Biol. 18, 68-76

HERZBERG, M., BREITBART, H., ATLAN, H. (1974) Interactions between membrane functions and protein synthesis in reticulocytes. Eur.J. Biochem. 45, 161-170

HETTINGER, T.P., CRAIG, L.C. (1970) Edeine. IV. Structures of the antibiotic peptides edeines A_1 and B_1. Biochemistry 9, 1223-1232

HICHENS, M., RINEHART, K.L. (1963) Chemistry of the neomycins. XII. The absolute configuration of deoxystreptamine in the neomycins, paromomycins and kanamycins. J.Amer.Chem.Soc. 85, 1547-1548

HICKEY, E.D., WEBER, L.A., BAGLIONI, C. (1976) Inhibition of initiation of protein synthesis by 7-methylguanosine-5'-monophosphate. Proc.Nat.Acad.Sci. 73, 19-23

HICKEY, E.D., WEBER, L.A., BAGLIONI, C. (1977) A relation between inhibition of protein synthesis and conformation of 5'-phosphorylated 7-methylguanosine derivatives. J.Mol.Biol. 109, 173-185

HICKEY, E.D., WEBER, L.A., BAGLIONI, C., KIM, C.H., SARMA, R.H. (1977) A relation between inhibition of protein synthesis and conformation of 5'-phosphorylated 7-methylguanosine derivatives. J.Mol.Biol. 169, 173-183

HIEROWSKI, M. (1965) Inhibition of protein synthesis by chlortetracycline in the E. coli in vitro system. Proc.Nat.Acad.Sci. 53, 594-599

HIGHLAND, J.H., HOWARD, G.A., GORDON, J. (1975) Binding of thiostrepton to the ribosomes of Escherichia coli: characterization and stoichiometry of binding. Eur.J.Biochem. 53, 313-318

HIGHLAND, J.H., HOWARD, G.A., OCHSNER, E., STOFFLER, G., HASENBANK, R., GORDON, J. (1975) Identification of a ribosomal protein necessary for thiostrepton binding to Escherichia coli ribosomes. J.Biol.Chem. 250, 1141-1145

HIGHLAND, J.H., SMITH, R.L., BURKA, E., GORDON, J. (1974) The effect of L-1-tosylamido-2-phenethyl-chloromethyl-ketone on the activity of prokaryote and eukaryote tRNA binding factors. FEBS Letters 39, 96-98

HIRSCH, J.D., MARTELO, O.J. (1976) Inhibition of rabbit reticulocyte protein kinases by hemin. Biochem.Biophys.Res.Commun. 71, 926-932

HISHIZAWA, T., PESTKA, S. (1971) Studies on the formation on transfer ribonucleic acid-ribosome complexes. XVII. The effect of tRNA on aminoacyl-oligonucleotide binding to ribosomes. Arch.Biochem. Biophys. 147, 624-631

HODGIN, L.A., HÖGENAUER, G. (1974) The mode of action of pleuromutilin derivatives. Effect on cell-free polypeptide synthesis. Eur.J. Biochem. 47, 527-533

HOEKSEMA, H. (1964) Celesticetin, IV. The structure of celesticetin. J.Amer.Chem.Soc. 86, 4224-4225

HOEKSEMA, H., BANNISTER, B., BIRKENMEYER, R.D., KAGAN, F., MAGERLEIN, B.J., MACFELLER, F.A., SCHROEDER, W., SLOMP, G., HERR, R.R. (1964) Chemical studies on lincomycin. I. The structure of lincomycin. J.Amer.Chem.Soc. 86, 4223-4224

HOERZ, W., McCARTY, K.S. (1970) Initiation of protein synthesis in a rabbit reticulocyte lysate system. Biochim.Biophys.Acta 228, 526-535

HOGAN, B.L. (1969) The effect of innibitors of protein synthesis on the level of ribosomal subunits in ascites cells. Biochim. Biophys.Acta 182, 264-266

HÖGENAUER, G. (1975) The mode of action of pleuromutilin derivatives. Location and properties of the pleuromutilin binding site on Escherichia coli ribosomes. Eur.J.Biochem. 52, 93-98

HÖGENAUER, G. (1978) The mode of action of derivatives of the antibiotic pleuromutilin. In "Antibiotics, Vol. V" (Hahn, F.E., ed.), Springer-Verlag:Berlin-Heidelberg-New York (in press)

HONJO, T., NISHIZUKA, Y., HAYAISHI, O. (1969) Adenosine diphosphoribosylation of aminoacyl transferase II by diphtheria toxin. Cold Spring Harb. Symp.Quant.Biol. 34, 603-608

HORGEN, P.A., GRIFFIN, D.H. (1971) Specific inhibitors of the three RNA polymerases from the aquatic fungus Blastocladiella emersonii. Proc.Nat.Acad.Sci. 68, 338-341

HUANG, K.T., MISATO, T., ASUYAMA, H. (1964) Effect of blasticidin S on protein synthesis of Piricularia oryzae. J.Antibiotics 17, 65-70

HUANG, M.T. (1976) Harringtonine, an inhibitor of initiation of protein biosynthesis. Mol.Pharmacol. 11, 511-519

HUANG, M.T., GROLLMAN, A.P. (1972a) Effects of aurintricarboxylic acid on ribosomes and the biosynthesis of globin in rabbit reticulocytes. Mol.Pharmacol. 8, 111-127

HUANG, M.T., GROLLMAN, A.P. (1972b) Mode of action of tylocrebrine: effects on protein and nucleic acid synthesis. Mol.Pharmacol. 8, 538-550

HUANG, M.T., GROLLMAN, A.P. (1973) Pyrocatechol violet: an inhibitor of initiation of protein synthesis. Biochem.Biophys.Res.Commun. 53, 1049-1095

HUNT, D.E., PITILLO, R.F., JOHNSON, E.P., MONCRIEF, F.C. (1966) The effect of actinobolin on nucleic acid and protein synthesis in Escherichia coli. Can.J.Microbiol. 12, 515-520

HURWITZ, C., BRAUN, C.B. (1968) Temperature-sensitivity of the weak bonds by which chloramphenicol is held in intact cells. Biochim.Biophys.Acta 157, 392-403

IBRAHIM, N.G., BEATTIE, D.S. (1973) Protein synthesis on ribosomes isolated from rat liver mitochondria: sensitivity to erythromycin. FEBS Letters 36, 102-104

IBRAHIM, N.G., BURKE, J.P., BEATTIE, D. (1974) The sensitivity of rat liver and yeast mitochondrial ribosomes to inhibitors of protein synthesis. J.Biol.Chem. 249, 6806-6811

IGARASHI, K., ISHITSUKA, H., KAJI, A. (1969) Comparative studies on the mechanism of action of lincomycin, streptomycin and erythromycin. Biochem.Biophys.Res.Commun. 37, 499-504

IGARASHI, S.J., ZMEAN, J.A. (1975) Effect of triphenylmethane derivatives of cell-free macromolecular synthesis. II. mRNA-ribosome binding. Canadian J.Biochem. 53, 124-127

IGLEWSKI, B.H., KABAT, D. (1975) NAD-dependent inhibition of protein synthesis by Pseudomonas aeruginosa toxin. Proc.Nat.Acad.Sci. 72, 2284-2288

IGLEWSKI, B.H., RITTENBERG, M.B. (1974) Selective toxicity of diphtheria toxin for malignant cells. Proc.Nat.Acad.Sci. 71, 2707-2710

IKEKAWA, T., UMEZAWA, H., SITAKA, Y. (1966) The structure of kasugamycin hydrobromide by X-ray crystallographic analysis. J.Antibiotics 19, 49-50

INOUE-YOKOSAWA, N., ISHIKAWA, C., KAZIRO, Y. (1974) The role of guanosine triphosphate in translocation reaction catalyzed by elongation factor G. J.Biol.Chem. 249, 4321-4323

ISHIDA, S., YAMADA, O., FUTATSUYA, F., ITO, K., YAMAMOTO, H., MUNAKATA, K. (1974) New herbicides derived from an antibiotic anysomycin. Proc.First Intersect. Congr. of IAMS. 3, 641-650

ITO,T.,MOTOHIRO,N.,OSAWA,H. (1964) The structure of kanamycin B. J.Antibiotics 17,189-193

ITOH, T., KOSUGI, H., HIGO, K., OSAWA, S. (1975) Ribosomal proteins from streptomycin-resistant and dependent mutants, and revertants from streptomycin-dependence to independence in Bacillus subtilis. Mol.Gen.Genetics 139, 293-301

ITOH, T., WITTMANN, H.G. (1973) Amino acid replacements in protein S5 and S12 of two Escherichia coli revertants from streptomycin dependence to independence. Mol.Gen.Genetics 127, 19-32

JACKSON, R.J. (1975) In "MTP International Review of Science. Biochemistry. Vol. 7. Synthesis of Amino Acids and Proteins" (Arnstein, H.R.V., ed.), pp. 89-135, Oxford:Medical and Technical Publishing Co

JACOBS-LORENA, M., BREGA, A., BAGLIONI, C. (1971) Inhibition of protein synthesis in reticulocytes by antibiotics. V. Mechanism of action of pederine, an inhibitor of initiation and elongation. Biochim.Biophys.Acta 240, 263-272

JAKES, K.S., ZINDER, N.D. (1974a) Highly purified colicin E3 contains immunity protein. Proc.Nat.Acad.Sci. 71, 3380-3384

JAKES, K.S., ZINDER, N.D. (1974b) Stimulation of in vitro colicin E3 action by messenger ribonucleic acid and message analogs. J.Biol.Chem. 249, 5507-5512

JAYARAMAN, J., GOLDBERG, I.H. (1968) Localization of sparsomycin action to the peptide-bond-forming step. Biochemistry 7, 418-421

JIMENEZ, A., CARRASCO, L., VAZQUEZ, D. (1977) Enzymic and non enzymic translocation by yeast polysomes. Site of action of a number of inhibitors. Biochemistry 16, 4727-4730

JIMENEZ, A., LITTLEWOOD, B., DAVIES, J. (1972) In "Molecular Mechanisms of Antibiotic Action on Protein Synthesis and Membranes" (Muñoz, E., García-Ferrandiz, F. and Vázquez, D., eds.), pp. 292-306, Amsterdam: Elsevier

JIMENEZ, A., MONRO, R.E., VAZQUEZ, D. (1970a) Interaction of Ac-Phe-tRNA with E. coli ribosomal subunit. Sparsomycin-induced formation of a complex containing 50S and 30S subunits but not mRNA. FEBS Letters 7, 103-108

JIMENEZ, A., MONRO, R.E., VAZQUEZ, D. (1970b) Interaction of Ac-Phe-tRNA with E. coli ribosomal subunit. Resistance of the sparsomycin-induced complex to hydroxylamine action. FEBS Letters 7, 109-113

JIMENEZ, A., SANTOS, A., ALONSO, G., VAZQUEZ, D. (1976) Inhibitors of protein synthesis in eukaryotic cells. Comparative effects of some Amaryllidaceae alkaloids. Biochim.Biophys.Acta 425, 342-348

JIMENEZ, A., SANCHEZ, L., VAZQUEZ, D. (1975a) Simultaneous ribosomal resistance to trichodermin and anisomycin in Saccharomyces cerevisiae mutants. Biochim.Biophys.Acta 383, 427-434

JIMENEZ, A., SANCHEZ, L., VAZQUEZ, D. (1975b) Yeast ribosomal sensitivity and resistance to the Amaryllidaceae alkaloids. FEBS Letters 60, 66-70

JIMENEZ, A., SANCHEZ, L., VAZQUEZ, D. (1975c) Location of resistance to the alkaloid narciclasine in the 60S ribosomal subunit. FEBS Letters 55, 53-56

JIMENEZ, A., VAZQUEZ, D. (1975) Quantitative binding of antibiotics to ribosomes from a yeast mutant altered on the peptidyl-transferase center. Eur.J.Biochem. 54, 483-492

JIMENEZ, A., VAZQUEZ, D. (1978) Anisomycin and related antibiotics. In "Antibiotics. Vol. V" (Hahn, F.E., ed.), Springer-Verlag: Berlin-Heidelberg-New York (in press)

JOHNSON, F., STARKOVSKY, N.A., PATON, A.C., CARLSON, A.A. (1964) Glutarimide antibiotics. IV. The total synthesis of dl- and l-cycloheximide. J.Amer.Chem.Soc. 86, 118-119

JONAK, J., CLARK, B.F.C. (1973) Inhibition of magic spot formation by L-1-tosylamido-2-phenylethyl-chloromethyl-ketone. FEBS Letters 34, 81-84

JONAK, J., SEDLACEK, J., RYCHLIK, I. (1971) Tosylphenylalanyl chloromethane-inhibitor of complex of S_1S_3-factors in cell-free protein-synthesizing system from Bacillus stearothermophilus. FEBS Letters 18, 6-8

JONAK, J., SEDLACEK, J., RYCHLIK, I. (1973) Mode of action of N-tosyl-L-phenylalanyl chloromethane on the elongation protein-synthesizing S_3 factor from Bacillus stearothermophilus. Biochim.Biophys.Acta 294, 322-328

JORDAN, D.C., YAMAMURA, Y., McKAGUE, M.E. (1969) Mode of action of viomycin on Rhizobium meliloti. Can.J.Microbiol. 15, 1005-1012

JOSTEN, J.J., ALLEN, P.M. (1964) The mode of action of lincomycin. Biochem.Biophys.Res.Commun. 14, 241-244

KAEMPFER, R., KAUFMAN, J. (1973) Inhibition of cellular protein synthesis by double-stranded RNA: inactivation of an initiation factor. Proc.Nat.Acad.Sci. 70, 1222-1226

KAJI, A. (1973) In "Progress in Molecular and Subcellular Biology" (Hahn, F.E., ed.), Vol. 3, pp. 85-158, Berlin-Heidelberg-New York: Springer-Verlag

KAJI, A., IGARASHI, K., ISHITSUKA, H. (1969) Interaction of tRNA with ribosomes-binding and release of tRNA. Cold Spring Harb. Symp.Quant.Biol. 34, 167-177

KAJI, H., TANAKA, Y. (1968) Binding of dihydrostreptomycin to ribosomal subunits. J.Mol.Biol. 32, 221-230

KALTSCHMIDT, E., WITTMANN, H.G. (1970) Ribosomal proteins. XII. Number of proteins in small and large ribosomal subunits of Escherichia coli as determined by two-dimensional gel electrophoresis. Proc.Nat.Acad.Sci. 67, 1276-1282

KANZAKI, T., HIGASHIDE, E., YAMAMOTO, H., SHIBATA, M., NAKAZAWA, K., IWASAKI, H., TAKEWAKA, T., MIYAKE, A. (1962) Gougerotin, a new antibacterial antibiotic. J.Antibiotics 15, 93-97

KAPPEN, L.S., GOLDBERG, I.H. (1973) Inhibition of globin chain initiation in reticulocyte lysates by pactamycin: accumulation of methionyl-valine. Biochem.Biophys.Res.Commun. 54, 1083-1091

KAPPEN, L.S., GOLDBERG, I.H. (1976) Analysis of the two steps in polypeptide chain initiation inhibited by pactamycin. Biochemistry 15, 811-818

KAPPEN, L.S., SUZUKI, H., GOLDBERG, I.H. (1973) Inhibition of reticulocyte peptide-chain initiation by pactamycin: accumulation of inactive ribosomal initiation complexes. Proc.Nat.Acad.Sci. 70, 22-26

KAUFMANN, Y., GOLDSTEIN, E., PENMAN, S. (1976) Poliovirus-induced inhibition of polypeptide initiation in vitro on native polyribosomes. Proc.Nat.Acad.Sci. 73, 1834-1838

KAUFMANN, Y., ZAMIR, A. (1973) Protection of E. coli ribosomes against colicin E3-induced inactivation by bound aminoacyl-tRNA. FEBS Letters 36, 277-280

KAUFMANN, Y., ZAMIR, A. (1975) The effect of tRNA derivatives bound with natural or synthetic mRNA on the interaction of Escherichia coli ribosomes with colicin E3. Eur.J.Biochem. 53, 599-603

KAVAGUCHI, H., NAITO, T., NAKAGAWA, S., FUJISAWA, K. (1972) BB-K8, a new semisynthetic aminoglycoside antibiotic. J.Antibiotics 25, 695-708

KAZIRO, Y. (1973) In "Organization of Energy-Transclucing Membranes" (Nakao, M. and Packer, L., eds.), pp. 187-200, Tokyo: University of Tokyo

KERR, I.M., BROWN, R.E. (1978) pppA2'p5'A2'p5'A: an inhibitor of protein synthesis synthesized with an enzyme fraction from interferon-treated cells. Proc.Nat.Acad.Sci. 75, 256-260

KERR, I.M., BROWN, R.E., BALL, L.A. (1974) Increased sensitivity of cell-free protein synthesis to double-stranded RNA after interferon treatment. Nature 250, 57-59

KERR, I., BROWN, R.E., HOVANESSIAN, A.G. (1977) Nature of inhibitor of cell-free protein synthesis formed in response to interferon and double-stranded RNA. Nature 268, 540-542

KERRIDGE, D. (1958) The effect of actidione and other antifungal agents on nucleic acid and protein synthesis in Saccharomyces carlsbergensis. J.Gen.Microbiol. 19, 497-506

KHOKHLOV, A.I., BLINOV, N.O. (1970) Antibiotics: chemistry and mode of action. FEBS Letters 11, 1-7

KINOSHITA, T., KAWANO, G., TANAKA, N. (1968) Association of fusidic acid sensitivity with G factor in a protein-synthesising system. Biochem.Biophys.Res.Commun. 33, 769-773

KINOSHITA, T., LION, Y.F., TANAKA, N. (1971) Inhibition by thiopeptin of ribosomal function associated with T and G factors. Biochem. Biophys.Res.Commun. 44, 859-863

KINOSHITA, T., TANAKA, N., UMEZAWA, H. (1970) Binding of blasticidin S to ribosomes. J.Antibiotics 23, 288-290

KLEIN, D.C., WELLER, J.L., KIRK, K.L., HARTLEY, R.W. (1977) Incorporation of 2-fluoro-L-histidine into cellular protein. Mol.Pharmacol. 13, 1105-1110

KO, T.S., KAJI, A. (1975) Effect of ricin on biosynthesis of myeloma protein (IgA) and general cellular proteins in MOPC-315 cells. Biochim.Biophys.Acta 414, 155-160

KOBAYASHI, H., KOBAYASHI, K., KOBAYASHI, Y. (1977) Isolation and characterization of fusidic acid-resistant, sporulation-defective mutants of Bacillus subtilis. J.Bacteriol. 132, 262-269

KOCH, K.F., RHOADES, J.A. (1968) Structure of nebramycin factor G, a new aminoglycoside antibiotic. Ant.Agents Chemother. 1967, 309-313

KOKA, M., NAKAMOTO, T. (1968) Studies on polyphenylalanine synthesis with a cell-free system of rat limphosarcoma. Biochim.Biophys. Acta 262, 381-392

KONDO, S., SHIBAHARA, S., TAKAHASHI, K., MAEDA, K., UMEZAWA, H. (1971) Negamycin, a novel hydrazide antibiotic. J.Amer.Chem.Soc. 93, 6305-6306

KONISHI, M., KIMEDA, M., TSUKIURA, H., YAMAMOTO, H., HOSHIYA, T., MIYAKI, T., FUJISAWA, K.-I., KOSHIYAMA, H., KAWAGUCHI, M. (1973) Oxamicetin, a new antibiotic of bacterial origin. I. Production, isolation and properties. J.Antibiotics 26, 752-756

KONISKY, J., NOMURA, M. (1967) Interaction of colicins with bacterial cells. II. Specific alteration of Escherichia coli ribosomes induced by colicin E3 in vivo. J.Mol.Biol. 26, 181-195

KORNFELD, E.C., JONES, R.G. (1948) The structure of actidione, an antibiotic from Streptomyces griseus. Science 108, 437-438

KORZYBSKI, T., KOWSZYK-GINDIFER, Z., KURYLOWICZ, W. (1967a) In "Antibiotics: Origin, Nature and Properties", Vol. I, Pergamon Press, Oxford

KORZYBSKI, T., KOWSZYK-GINDIFER, Z., KURYLOWICZ, W. (1967b) In "Antibiotics: Origin, Nature and Properties", Vol. II, Pergamon Press, Oxford

KOZAK, M., NATHANS, D. (1972) Differential inhibition of coliphage MS2 protein synthesis by ribosome-directed antibiotics. J.Mol. Biol. 70, 41-55

KRAMER, G., CIMADEVILLA, J.M., HARDESTY, B. (1976) Specificity of the protein kinase activity associated with the hemin-controlled repressor of rabbit reticulocyte. Proc.Nat.Acad.Sci. 73, 3078-3082

KREDICH, N.M., GUARINO, A.J. (1961) Homocitrullylaminoadenosine, a nucleoside isolated from Cordyceps militaris. J.Biol.Chem. 236, 3300-3302

KROON, A.M., DeVRIES, H. (1970) In "Control of Organelle Development" (Miller, P.L., ed.), pp. 181-199, Cambridge: Cambridge University Press

KROON, A.M., DeVRIES, H. (1971) In "Autonomy and Biogenesis of Mitochondria and Chloroplasts" (Boardman, N.K., Linnane, A.W. and Smillie, R.M., eds.), Amsterdam: North Holland

KUBOTA, K., OKUYAMA, A., TANAKA, N. (1972) Differential effects of antibiotics on peptidyl transferase reactions. Biochem.Biophys. Res.Commun. 47, 1196-1202

KUCAN, Z., LIPMANN, F. (1964) Differences in chloramphenicol sensitivity of cell-free amino acid polymerization systems. J.Biol.Chem. 239, 516-520

KURYLO-BOROWSKA, Z. (1962) On the mode of action of edeine. Biochim. Biophys.Acta 61, 897-902

KURYLO-BOROWSKA, Z. (1964) On the mode of action of edeine. Biochim. Biophys.Acta 87, 305-313

KURYLO-BOROWSKA, Z., SZER, W. (1971) Inhibition of bacterial DNA synthesis by edeine. Effect on Escherichia coli mutants lacking DNA polymerase I. Biochim.Biophys.Acta 287, 236-245

KYOICHI, A., WATANABE, S., FALCO, J., FOX, J.J. (1972) Total synthesis of gougerotin. J.Amer.Chem.Soc. 94, 3272-3274

LAI, C.J., WEISBLUM, B. (1971) Altered methylation of ribosomal RNA in an erythromycin-resistant strain of Staphylococcus aureus. Proc.Nat.Acad.Sci. 68, 856-860

LAI, C.J., WEISBLUM, B., FAHNESTOCK, S.R., NOMURA, M. (1973) Alteration of 23S ribosomal RNA and erythromycin-induced resistance to lincomycin and spiramycin in Staphylococcus aureus. J.Mol.Biol. 74, 67-72

LANDO, D., COUSIN, M.A., GARILHE, M.PRIVAT de. (1973) Misreading, a fundamental aspect of the mechanism of action of several aminoglycosides. Biochemistry 12, 4528-4533

LANDO, D., COUSIN, M.A., OJASOO, T., RAYNAUD, J.P. (1976) Paromomycin and dihydrostreptomycin binding to Escherichia coli ribosomes. Eur.J.Biochem. 66, 597-606

LASKIN, A.I. (1967) In "Antibiotics. Vol. I. Mechanism of Action" (Gottlieb, D. and Shaw, P.D., eds.), pp. 331-359, Springer-Verlag, Berlin, Heidelberg, New York

LASKIN, A.I., CHAN, W.M. (1965) Ant.Agents Chemother. 1964, 485-490

LAU, C., RICHARDS, F.M. (1976) Proteolytic and chemical modification of colicin E3 activity. Biochemistry 15, 3856-3863

LAWRENCE, C., THACH, R.E. (1974) Encephalomyocarditis virus infection of mouse plasmacytoma cells. I. Inhibition of cellular protein synthesis. J.Virol. 14, 598-610

LAZAR, M., GROS, F. (1973) Translation initiation defects in ribosomes from streptomycin dependent strains. Biochimie 55, 171-181

LEACH, B.E., CALHOUN, K.M., JOHNSON, L.E., TEETERS, C.M., JACKSON, W.C. (1953) Chartreusin, a new antibiotic produced by Streptomyces chartreusis, a new species. J.Amer.Chem.Soc. 75, 4011-4012

LEADER, D.P., RANKINE, A.D., COIA, A.A. (1976) The phosphorylation of ribosomal protein S6 in baby hamster Kidney fibroblast. Biochem.Biophys.Res.Commun. 71, 966-774

LEBLEU, B., MARBAIX, G., WERENNE, J., BURNY, A., HUEZ, G. (1970) Effect of aurintricarboxylic acid and of NaF on the binding of globin messenger RNA to reticulocyte 40S ribosomal subunits. Biochem.Biophys.Res.Commun. 40, 731-739

LEBLEU, B., SEN, G.C., SHAILA, S., CABRER, B., LENGYEL, P. (1976) Interferon, double-stranded RNA,and protein phosphorylation. Proc.Nat.Acad.Sci. 73, 3107-3111

LEE-HUANG, S., SIERRA, J.M., NARANJO, R., FILIPOWICZ, W., OCHOA, S. (1977) Eukaryotic oligonucleotides affecting mRNA translation. Arch.Biochem.Biophys. 180, 276-287

LELONG, J.C., COUSIN, M.A., GROS, D., GRUNBERG-MANAGO, M., GROS, F. (1971) Streptomycin induced release of fMet-tRNA from the ribosomal initiation complex. Biochem.Biophys.Res.Commun. 42, 530-537

LELONG, J.C., COUSIN, H.A., GROS, F., MISKIN, R., VOGEL, Z., GRONER, Y., REVEL, M. (1972) Protection of Escherichia coli ribosomes against streptomycin by purified initiation factors. Eur.J.Biochem. 27, 174-180

LELONG, J.C., GROS, D., GROS, F., BOLLEN, A., MASCHLER, R., STOFFLER, G. (1974) Function of individual 30S subunit proteins of Escherichia coli. Effect of specific immunoglobulin fragments on activities of ribosomal decoding sites. Proc.Nat.Acad.Sci. 71, 248-252

LEMIEUX, R.U., WOLFROM, M.L. (1948) The chemistry of streptomycin. Adv.Carbohydrate Chemistry 3, 337-384

LESSARD, J.L., PESTKA, S. (1972) Studies on the formation of transfer ribonucleic acid-ribosome complexes. XXIII. Chloramphenicol, aminoacyl-oligonucleotides, and Escherichia coli ribosomes. J.Biol. Chem. 247, 6909-6912

LEVIN, D.H., KYNER, D., ACS, G. (1973) Protein synthesis initiation in eukaryotes. Characterization of ribosomal factors from mouse fibroblasts. J.Biol.Chem. 248, 6416-6425

LEVIN, D.H., RANU, R.S., ERNST, V., LONDON, I.M. (1976) Regulation of protein synthesis in reticulocyte lysates: phosphorylation of methionyl-$tRNA_f$ binding factor by protein kinase activity of translational inhibitor isolated from heme-deficient lysates. Proc.Nat.Acad.Sci. 73, 3112-3116

LEWIS, C. (1974) Clinically useful antibiotics obtained by directed chemical modification-lincomycins. Fed.Proceedings 33, 2303-2306

LIAO, L.L., GROLLMAN, A.P., HORWITZ, S.B. (1976) Mechanism of action of the 12,13 epoxytrichothecene, anguidine, an inhibitor of protein synthesis. Biochim.Biophys.Acta 454, 273-284

LIAO, L.L., KUPCHAN, S.M., HORWITZ, S.B. (1976) Mode of action of the antitumor compound bruceantin, an inhibitor of protein synthesis. Mol.Pharmacol. 12, 167-176

LICHTENTHALER, F.W., CERNA, J., RYCHLIK, I. (1975) The effect of oxamicetin and some amicetin analogs on ribosomal peptidyl transferase. FEBS Letters 53, 184-187

LIETMAN, P.S. (1971) Mitochondrial protein synthesis: inhibition by emetine hydrochloride. Mol.Pharmacol. 7, 122-128

LIN, J.Y., KAO, W.Y., TSERNG, K.Y., CHEN, C.C., TUNG, T.C. (1970a) Effect of crystalline abrin on the biosynthesis of protein, RNA and DNA in experimental tumors. Cancer Res. 30, 2431-2433

LIN, J.Y., LIU, K., CHEN, C.C., TUNG, T.C. (1971) Effect of crystalline ricin on the biosynthesis of protein, RNA and DNA in experimental tumors. Cancer Res. 31, 921-923

LIN, J.Y., PAO, C.C., JU, S.T., TUNG, T.C. (1972) Polyribosome disaggregation in rat liver following administration of the phytotoxic proteins, abrin and ricin. Cancer Res. 32, 943-947

LIN, J.Y., TSERNG, K.Y., CHEN, C.C., LIN, L.T., TUNG, T.C. (1970b) Abrin and ricin: new antitumour substances. Nature 237, 292-293

LIN, Y.C., KINOSHITA, T., TANAKA, N. (1968) Mechanism of protein synthesis inhibition by bottromycin A_2: studies with puromycin. J.Antibiotics 21, 471-476

LIN, Y.C., TANAKA, N. (1968) Mechanism of action of bottromycin in polypeptide biosynthesis. J.Biochem. 63, 1-7

LIN, Y.C., TANAKA, N., UMEZAWA, H. (1967) Bottromycin action on macromolecular biosynthesis. J.Antibiotics 20, 223-226

LIOU, Y.F., KINOSHITA, T., TANAKA, N., YOSHIKAWA, M. (1973) Studies on the ribosomes of a thiopeptin-resistant mutant of Escherichia coli. J.Antibiotics 26, 711-716

LIOU, Y.F., TANAKA, N. (1976) Dual actions of viomycin on the ribosomal functions. Biochem.Biophys.Res.Commun. 71, 477-483

LIOU, Y.F., YOSHIKAWA, M., TANAKA, N. (1975) Alteration of ribosomal protein L5 in a thiopeptin-resistant mutant of Escherichia coli. Biochem.Biophys.Res.Commun. 65, 1096-1101

LOCKWOOD, A.H., SARKAR, P., MAITRA, U., BROT, N., WEISSBACH, H. (1974) Effect of thiostrepton on polypeptide chain initiation in Escherichia coli. J.Biol.Chem. 249, 5831-5834

LODISH, H.F., HOUSSMAN, D., JACOBSEN, M. (1971) Initiation of hemoglobin synthesis. Specific inhibition by antibiotics and bacteriophage ribonucleic acid. Biochemistry 10, 2348-2356

LODISH, H.F., ROSE, J.K. (1977) Relative importance of 7-methyl-guanosine in ribosome binding and translation of vesicular stomatitis virus mRNA in wheat germ and reticulocyte cell-free systems. J.Biol.Chem. 252, 1181-1188

LOEB, J.E., BLAT, C. (1970) Phosphorylation of some rat liver ribosomal proteins and its activation by cyclie AMP. FEBS Letters 10, 105-108

LUCAS-LENARD, J., HAENNI, A.L. (1968) Requirement of guanosine 5'-triphosphate for ribosomal binding of aminoacyl-SRNA. Proc.Nat. Acad.Sci. 59, 554-560

LUCAS-LENARD, J., LIPMANN, F. (1971) Protein biosynthesis. Ann.Rev. Biochem. 40, 409-448

LUGNIER, A., DIRHEIMER, G. (1973) Differences between ricin and phytohemagglutinins from Ricinus communis seeds. FEBS Letters 35, 117-120

LUGNIER, A.A.J., KUNTZEL, H., DIRHEIMER, G. (1976) Inhibition of Neurospora crassa and yeast mitochondrial protein synthesis by ricin, a toxic protein inactive on E. coli protein synthesis. FEBS Letters 66, 202-205

LUGNIER, A.A.J., LE MEUR, M.A., GERLINGER, P., DIRHEIMER, G. (1974) Inhibition of in vitro protein synthesis in a rabbit reticulocyte cell-free system by toxic trypsic peptides from ricin. Biochimie 56, 1287-1289

LUZZATTO, L., APIRION, D., SCHLESSINGER, D. (1968) Mechanism of action of streptomycin in E. coli: interruption of the ribosome cycle at the initiation of protein synthesis. Proc.Nat.Acad.Sci. 60, 873-880

LUZZATTO, L., APIRION, D., SCHLESSINGER, D. (1969) Polyribosome depletion and blockage of the ribosome cycle by streptomycin in Escherichia coli. J.Mol.Biol. 42, 315-335

MacDONALD, J.S., GOLDBERG, I.H. (1970) An effect of pactamycin on the initiation of protein synthesis in reticulocytes. Biochem. Biophys.Res.Commun. 41, 1-8

MADEN, B.E.H., MONRO, R.E. (1968) Ribosome-catalyzed peptidyl transfer: the polyphenylalanine synthesis. Eur.J.Biochem. 6, 309-316

MAEHR, H., LEACH, M., YARMCHUK, L., STEMPEL, A. (1973) Antibiotic X-5108. V. Structures of antibiotic X-5108 and mocimycin. J.Amer. Chem.Soc 95, 8449-8450

MAEHR, H., SHAFFNER, J.P. (1970) Chemistry of the gentamicins. II. Stereochemistry and synthesis of gentosamine. Total structure of gentamicin A. J.Amer.Chem.Soc. 92, 1697-1700

MAGER, J., BENEDICT, M., ARTMAN, M. (1962) A common site of action for polyamines and streptomycin. Biochim.Biophys.Acta 62, 202-204

MALKIN, N., LIPMANN, F. (1969) Fusidic acid: inhibition of factor T2 in reticulocyte protein synthesis. Science 164, 71-72

MAO, J.C.H., PUTTERMAN, M. (1969) The intermolecular complex of erythromycin and ribosome. J.Mol.Biol. 44, 347-361

MAO, J.C.H., ROBISHAW, E.E. (1971) Effects of macrolides on peptide bond formation and translocation. Biochemistry 10, 2054-2061

MAO, J.C.H., ROBISHAW, E.E. (1972) Erythromycin a peptidyl-transferase effector. Biochemistry 11, 4864-4872

MARCUS, A., BEWLEY, J.D., WEEKS, D.P. (1970) Aurintricarboxylic acid and initiation factors of wheat embryo. Science 167, 1735-1736

MARCUS, A., FEELEY, J. (1966) Ribosome activation and polysome formation in vitro: requirement for ATP. Proc.Nat.Acad.Sci. 56, 1770-1777

MARGULIES, L.J., KAY, J.E. (1976) The mechanism of inhibition of initiation of protein synthesis in reticulocyte lysates by pyrocatechol violet. Biochim.Biophys.Acta 435, 152-158

MARKS, P.A., BURKA, E.R., CONCONI, F., PERL, W., RIFKIND, R.A. (1965) Polyribosomes dissociation and formation in intact reticulocytes with conservation of messenger ribonucleic acid. Proc.Nat.Acad. Sci. 53, 1437-1443

MARSH, R.C., PARMEGGIANI, A. (1973) Requirement of proteins S5 and S9 from 30S subunit for the ribosome-dependent GTPase activity of elongation factor G. CsCl 30S core/ribosomal split proteins/30S reconstitution/30S-50S association. Proc.Nat.Acad.Sci. 70, 151-155

MARTINEZ, O., VAZQUEZ, D., MODOLELL, J. (1978) Streptomycin- and viomycin-induced conformational changes of ribosomes detected by iodination. FEBS Letters 87, 21-28

MARYANKA, D., JOHNSTON, I.R. (1970) The inhibition of DNA-dependent RNA polymerase of E. coli by showdomycin. FEBS Letters 7, 125-128

MASON, D.J., DIETZ, A., DEBOER, C. (1963) Lincomycin, a new antibiotic. I. Discovery and biological properties. Ant.Agents Chemother.- 1962, 554-559

MASON, D.J., DIETZ, A., SMITH, R.M. (1961) Actinospectacin, a new antibiotic. I. Discovery and biological properties. Antibiotics Chemother. 11, 118-122

MASUKAWA, H., TANAKA, N. (1968) Miscoding activity of amino sugars. J.Antibiotics 21, 70-72

MASUKAWA, H., TANAKA, N., UMEZAWA, H. (1968) Inhibition by kasugamycin of protein synthesis in Piricularia oryzae. J.Antibiotics 21, 73-74

MATHEWS, M.B., KORNER, A. (1970) Mammalian cell-free protein synthesis directed by viral ribonucleic acid. Eur.J.Biochem. 17, 328-338

MAXWELL, E.S., ROBINSON, E.A., HENRIKSEN, O. (1975) Elongation factor 2: amino acid sequence at the site of ADP-ribosylation. J.Biochem. 77, 9

MAXWELL, I.H. (1966) Studies on the binding of tetracycline to ribosomes in vitro. Mol.Pharmacol. 4, 25-37

MAZUMDER, R. (1973) Effect of thiostrepton on recycling of Escherichia coli initiation factor 2. Proc.Nat.Acad.Sci. 70, 1939-1942

McCORMICK, M.H., HOEHN, M.M. (1953) Isolation of a new antibiotic from Streptomyces fasciculatus nov.sp. Antibiotics Chemother. 3, 718-720

McCORMICK, L., PENMAN, S. (1969) Regulation of protein synthesis in HeLa cells: translation at elevated temperatures. J.Mol.Biol. 39, 315-333

McPHERSON, A., RICH, A. (1973) Studies on crystalline abrin: X-ray diffraction data, molecular weight, carbohydrate content and subunit structure. FEBS Letters 35, 257-261

MEAGER, A., UNGKITCHANUKIT, A., HUGHES, R.C. (1976) Variants of hamster fibroblasts to Ricinus communis toxin (ricin). Biochem. J., 154, 113-124

MENNINGER, J.R. (1971) A simple assay for protein chain termination using natural peptidyl-tRNA. Biochim.Biophys.Acta 240, 237-243

MESZAROS, K., MANDL, J., ANTONI, F., GARZO, T. (1976) Inhibition of protein synthesis by hexosamine containing glycogen formed in mouse liver after treatment with Dugalactosamine. FEBS Letters 71, 215-219

MILLER, D.L. (1972) Elongation factors EF Tu and EF G interact at related sites on ribosomes. Proc.Nat.Acad.Sci. 69, 752-755

MINKS, M.A., ARIATTI, M., HAWTREY, A.O. (1975) Studies on the binding of N-bromoacetylpuromycin and N-bromoacetylaminonucleoside to rat liver ribosomes. Hoppe-Seyler's Z.Physiol.Chem. 356, 109-118

MISKIN, R., ZAMIR, A. (1972) Effect of streptomycin on ribosome interconversion, a possible basis for the action of the antibiotic. Nature New Biol. 238, 78-80

MISUMI, M., TANAKA, N., SHIBA, T. (1978) Binding of (^{14}C) tuberactinomycin O, an antibiotic closely related to viomycin, to the bacterial ribosome. Biochem. Biophys. Res. Commun. 82, 971-976

MITCHELL, S.J., HEDBLOM, M., CAWLEY, D., HOUSTON, L.L. (1976) Ricin does not act as an endonuclease on L cells polysomal RNA. Biochem.Biophys.Res.Commun. 68, 763-769

MITSUHASHI, S. (ed.) (1977) R Factor Drug Resistance Plasmid. University Park Press, Baltimore, Maryland, USA

MITSUHASHI, S., HASHIMOTO, H. (eds.) (1977) Microbial Drug Resistance. University Park Press. Baltimore, Maryland, USA

MITSUHASHI, S., ROSIVAL, L., KRCMERY, V. (eds.) (1975) Drug-Inactivating Enzymes and Antibiotic Resistance. Springer-Verlag, Berlin-Heidelberg-New York

MIYAIRI, N., MIYOSHI, T., AOKI, H., KOHSAKA, M., IKUSHINA, H., KUNUGITA, K., SAKAI, H., IMANAKA, H. (1970) Studies on thiopeptin antibiotics. I. Characteristics of thiopeptin B. J.Antibiotics 23, 113-119

MIYAZAWA, F., OLIJNYK, O.B., TILLEY, C.J., TAMAOKI, T. (1967) Interaction between dextran sulphate and *Escherichia coli* ribosomes. Biochim.Biophys.Acta 145, 96-104

MIZUNO, S. (1975) Mechanism of inhibition of protein synthesis initiation by diacetoxyscirpenol and fusarenon X in the reticulocyte lysate system. Biochim.Biophys.Acta 383, 207-214

MIZUNO, S., NITTA, K., UMEZAWA, H. (1967a) Mode of action of enomycin, an antitumor antibiotic of high molecular weight. I. Inhibition of protein synthesis. J.Biochem. 61, 373-381

MIZUNO, S., NITTA, K., UMEZAWA, H. (1967b) Mode of action of enomycin, an antitumor antibiotic of high molecular weight. II. Preferential binding of enomycin to mammalian ribosomes. J.Biochem. 61, 382-387

MIZUNO, S., NITTA, K., UMEZAWA, H. (1970a) Mechanism of action of negamycin in Escherichia coli K12. I. Inhibition of initiation of protein synthesis. J.Antibiotics 23, 581-588

MIZUNO, S., NITTA, K., UMEZAWA, H. (1970b) Mechanism of action of negamycin in Escherichia coli K12. Miscodin activity in polypeptide synthesis directed by synthetic polynucleotide. J.Antibiotics 23, 589-594

MIZUNO, S., UMEZAWA, H. (1976) Inhibition of the initial dipeptide synthesis of globin chains by the antibiotic enomycin in the reticulocyte lysate. J.Antibiotics 29, 390-315

MODOLELL, J., CABRER, B., PARMEGGIANI, A., VAZQUEZ, D. (1971) Inhibition by siomycin and thiostrepton of both aminoacyl-tRNA and factor G binding to ribosomes. Proc.Nat.Acad.Sci. 68, 1796-1800

MODOLELL, J., CABRER, B., VAZQUEZ, D. (1973a) Elongation factor T-dependent hydrolysis of guanosine triphosphate resistant to thiostrepton. J.Biol.Chem. 248, 8356-8360

MODOLELL, J., CABRER, B., VAZQUEZ, D. (1973b) The stoichiometry of ribosomal translocation. J.Biol.Chem. 248, 8356-8360

MODOLELL, J., DAVIS, B.D. (1968) Rapid inhibition of polypeptide chain extension by streptomycin. Proc.Nat.Acad.Sci. 61, 1279-1286

MODOLELL, J., DAVIS, B.D. (1969) Mechanism of inhibition of ribosomes by streptomycin. Nature 224, 345-348

MODOLELL, J., DAVIS, B.D. (1970a) Breakdown by streptomycin of initiation complexes formed on ribosomes of E. coli. Proc.Nat. Acad.Sci. 67, 1148-1155

MODOLELL, J., DAVIS, B.D. (1970b) Significance of the effect of streptomycin and fusidic acid on polysome stability. In "Progress in Antimicrobial and Anticancer Chemotherapy", Vol. II, pp. 464-467, Tokyo: University of Tokyo Press

MODOLELL, J., GIRBES, T., VAZQUEZ, D. (1975) Ribosomal translocation promoted by guanylylimido diphosphate and guanylyl-methylene diphosphonate. FEBS Letters 60, 109-113

MODOLELL, J., VAZQUEZ, D. (1975) In "MTP International Review of Science. Biochemistry. Vol. 7. Synthesis of Amino Acids and Proteins" (Arnstein, H.R., ed.), pp. 137-178, Oxford: Medical and Technical Publishing Co

MODOLELL, J., VAZQUEZ, D. (1977) The inhibition of ribosomal translocation by viomycin. Eur.J.Biochem. 81, 491-497

MODOLELL, J., VAZQUEZ, D., MONRO, R.E. (1971) Ribosomes, G-factor, and siomycin. Nature New Biol. 230, 109-112

MÖLLER, W. (1974) In "Ribosomes" (Nomura, M., Tissieres, A. and Lengyel, P., eds.), pp. 711-731, New York: Cold Spring Harbor Laboratory, Cold Spring Harbor

MONDON, A., KROHN, K. (1975) Zur Kenntnis des Narciclasins. Chem.Ber. 108, 445-463

MONRO, R.E. (1967) Catalysis of peptide bond formation by 50S ribosomal subunits from Escherichia coli. J.Mol.Biol. 26, 147-151

MONRO, R.E., CELMA, M.L., VAZQUEZ, D. (1969) Action of sparsomycin on ribosome-catalyzed peptidyl transfer. Nature 222, 356-358

MONRO, R.E., FERNANDEZ-MUÑOZ, R., CELMA, M.L., JIMENEZ, A., BATTANER, E., VÁZQUEZ, D. (1970) Antibiotics acting on the peptidyl transferase center of ribosomes. In "Progress in Antimicrobial and Anticancer Chemotherapy", Vol. II, pp. 473-481, University of Tokyo Press, Tokyo

MONRO, R.E., FERNANDEZ-MUÑOZ, R., CELMA, M.L., VAZQUEZ, D. (1971) Mode of action of lincomycin and related antibiotics. In "Drug Action and Drug Resistance in Bacteria. 1. Macrolide Antibiotics and Lincomycin" (Mitsuhashi, S., ed.), pp. 305-336, University of Tokyo Press, Tokyo

MONRO, R.E., MADEN, B.E., TRAUT, R.R. (1967) In "The Genetic Elements: Properties and Function" (Shugar, D., ed.), pp. 179-203, Academic Press, London-New York

MONRO, R.E., STAEHELIN, T., CELMA, M.L., VAZQUEZ, D. (1969) The peptidyl transferase activity of ribosomes. Cold Spring Harb. Symp.Quant.Biol. 34, 357-368

MONRO, R.E., VAZQUEZ, D. (1967) Ribosome-catalysed peptidyl transfer: effects of some inhibitors of protein synthesis. J.Mol.Biol. 28, 161-165

MONTANARO, L., SPERTI, S., MATTIOLI, A., TESTONI, G., STIRPE, F. (1975) Inhibition by ricin of protein synthesis in vitro. Biochem.J. 146, 127-131

MONTANARO, L., SPERTI, S., STIRPE, F. (1973) Inhibition by ricin of protein synthesis in vitro. Ribosomes as the target of the toxin. Biochem.J. 136, 677-683

MONTANARO, L., SPERTI, S., TESTONI, G., MATTIOLI, A. (1976) Effect of elongation factor 2 and of adenosine diphosphate-ribosylated elongation factor 2 on translocation. Biochem.J. 156, 15-23

MONTANARO, L., SPERTI, S., ZAMBONI, M., DENARO, M., TESTONI, G., GASPERI-CAMPANI, A., STIRPE, F. (1978) Effect of modeccin on the steps of peptide chain elongation. Biochem.J. (in press)

MOORE, V.G., ATCHISON, R.E., THOMAS, G., MORAN, M.B., NOLLER, H.F. (1975) Identification of ribosomal protein essential for peptidyl transferase activity. Proc.Nat.Acad.Sci. 72, 844-848

MORI, T., ICHIYANAGI, T., KONDO, H., TOKUNAGA, K., ODA, T. (1971) Studies on the new antibiotic lividomycins. II. Isolation and characterization of lividomycins A, B and other aminoglycoside antibiotics produced by Streptomyces lividus. J.Antibiotics 24, 339-349

MOSKOWITZ, M.A., WEISS, B.F., LYTLE, L.D., MUNRO, H.N., WURTMAN, R.J. (1975) d-Amphetamine disaggregates brain polysomes via a dopaminergic mechanism. Proc.Nat.Acad.Sci. 72, 834-836

MOULE, Y., JEMMALI, M., DARRACQ, N. (1978) Inhibition of protein synthesis by PR toxin, a mycotoxin from Penicillium roqueforti. FEBS Letters 88, 341-344

MUNK, M.E., NELSON, D.B., ANTOSZ, F.J., HERALD, D.L., Jr., HASKELL, T.H. (1968) The structure of actinobolin. J.Amer.Chem.Soc. 90, 1087-1089

MUNRO, H.N., BALIGA, B.S., PRONCZUCK, A.W. (1968) In vitro inhibition of peptide synthesis and GTP hydrolysis by cycloheximide and reversal of inhibition by glutathione. Nature 219, 944-946

MUÑOZ, E., GARCIA-FERRANDIZ, F., VAZQUEZ, D. (eds.) (1972) Molecular Mechanism of Antibiotic Action of Protein Synthesis and Membranes. Elsevier, Amsterdam

MURASE, M. (1961) Structural studies on kanamycin C. J.Antibiotics 14, 367-368

NAAKTGEBOREN, N., VERMAAS, A., VOORMA, H.O. (1975) The joining of the 30S initiation complex with the 50S subunit the main target for thiostrepton. Eur.J.Biochem. 57, 493-501

NASHIMOTO, H., UCHIDA, H. (1975) Late steps in the assembly of 30S ribosomal proteins in vivo in a spectinomycin-resistant mutant of Escherichia coli. J.Mol.Biol. 96, 443-453

NATHANS, D., NEIDLE, A. (1963) Structural requirements for puromycin inhibition of protein synthesis. Nature 197, 1076-1077

NETH, R., MONRO, R.E., HELLER, C., BATTANER, E., VAZQUEZ, D. (1970) Catalysis of peptidyl transfer by human ribosomes and effects of some antibiotics. FEBS Letters _6_, 198-202

NEUSS, N., KOCH, K.F., MOLLOY, B.B., DAY, W., HUCKSTEP, L.L., DORMAN, D.E., ROBERTS, J.D. (1970) Structure of hygromycin B, an antibiotic from _Streptomyces hygroscopicus_; the use of CMR. Spectra in structure determination 1. Helv.Chim.Acta _53_, 2314-2319

NEWTON, B.A. (1965) Mechanisms of antibiotic action. Ann.Rev.Microbiol. _19_, 209-240

NEWTON, B.A., REYNOLDS, P.E. (eds.) (1966) Sixteenth Symposium of the Society for General Microbiology. Cambridge: Cambridge University Press

NICOLSON, G.L., BLAUSTEIN, J., ETZLER, M.E. (1974) Characterization of two plant lectins from _Ricinus communis_ and their quantitative interaction with a murine lymphoma. Biochemistry _13_, 196-204

NICOLSON, G.L., LACORBIERE, M., HUNTER, T.R. (1975) Mechanism of cell entry end toxicity of an affinity-purified lectin from _Ricinus communis_ and its differential effects on normal and virus-transformed fibroblasts. Cancer Res. _35_, 144-155

NIERHAUS, D., NIERHAUS, K.H. (1973) Identification of the chloramphenicol-binding protein in _Escherichia coli_ ribosomes by partial reconstitution. Proc.Nat.Acad.Sci. _70_, 2224-2228.

NIRENBERG, M.W., MATTHAEI, J.H. (1961) The dependence of cell-free protein synthesis upon naturally occurring or synthetic polynucleotides. Proc.Nat.Acad.Sci. _47_, 1588-1602

NISHIMURA, H., MAYAMA, M., KOMATSU, Y.,KATO, H., SHIMAOKA, N., TANAKA, Y. (1964) Showdomycin, a new antibiotic from a Streptomyces sp. J.Antibiotics _17_, 148-155

NISHIMURA, H., OKAMOTO, S., MAYAMA, M., OHTSUKA, H., NAKAJIMA, K., TAWARA, K., SHIMOHIRA, M., SHIMAOKA, N. (1961) Siomycin, a new thiostrepton-like antibiotic. J.Antibiotics 14, 255-263

NISHIMURA, T. (1968a) Mechanism of action of phenomycin, a tumor-inhibitory polypeptide. J.Antibiotics 21, 110-118

NISHIMURA, T. (1968b) Mechanism of action of phenomycin: comparative study with diphtheria toxin. J.Antibiotics 21, 519-520

NOLAN, R.D., GRASMUK, H., DREWS, J. (1976) The binding of tritiated elongation-factors 1 and 2 to ribosomes from Krebs II mouse ascites-tumor cells. The influence of various antibiotics and toxins. Eur.J.Biochem. 64, 69-75

NOMOTO, A., SATOH, H., UKITA, T. (1975) Effect of Ricinus communis toxin on cyclic adenosine 3': 5'-monophosphate metabolism in Yoshida ascites sarcoma cells. J.Biol.Chem. 250, 2703-2708

NOMOTO, S., TESHIMA, T., WAKAMIYA, T., SHIBA, T. (1977) The revised structure of capreomycin. J.Antibiotics 30, 955-959

NOMURA, M. (1976) Quoted by Chang, Ch. and Craven, G.R. (1977) Identification of several proteins involved in the messenger RNA binding site of the 30S ribosome by inactivation with 2-methoxy-5-nitrotropone. J.Mol.Biol. 117, 401-418

NOWAK, T.S., MUNRO, H.N. (1977) Inhibition of cell-free protein synthesis initiation by amphetamine: association with reduction in tRNA aminoacylation. Biochem.Biophys.Res.Commun. 77, 1280-1285

NUSS, D.L., KOCH, G. (1976) Translation of individual host mRNA's in MPC-11 cells is differentially suppressed after infection by vesicular stomatitis virus. J.Virol. 19, 572-578

OBRIG, T.G., CULP, W.J., McKEEHAN, W.L., HARDESTY, B. (1971) The mechanism by which cycloheximide and related glutarimide antibiotics inhibit peptide synthesis on reticulocyte ribosomes. J.Biol.Chem. 246, 174-181

OBRIG, T.G., IRVIN, J., CULF, W., HARDESTY, B. (1976) Inhibition of peptide initiation or reticulocyte ribosomes by edeine. Eur.J. Biochem. 21, 31-41

OBRIG, T.G., IRVIN, J.D., HARDESTY, B. (1973) The effect of an antiviral peptide on the ribosomal reactions of the peptide elongation enzymes, EF-I and EF-II. Arch.Biochem.Biophys. 155, 278-289

ODA, T., MORI, T., KYOTANI, Y., NAKAYAMA, M. (1971) Studies on new antibiotic lividomycins. IV. Structure of lividomycin A. J.Antibiotics 24, 511-518

OHNO, S., OHNO-IWASHITA, Y., SUZUKI, K., IMAHORI, K. (1977) Purification and characterization of active component and active fragment of colicin E3. J.Biochem. 82, 1045-1053

OHNO-IWASHITA, Y., IMAHORI, K. (1977) Colicin E3 induced cleavage of 16S rRNA of isolated 30S ribosomal subunits. J.Biochem. 82, 919-922

OHSUMI, Y., IMAHORI, K. (1974) Studies on a factor enhancing colicin E3 activity in vitro. Proc.Nat.Acad.Sci. 71, 4062-4066

OKURA, A., KINOSHITA, T., TANAKA, N. (1970) Complex formation of fusidic acid with G factor, ribosome and guanosine nucleotide. Biochem.Biophys.Res.Commun. 41, 1545-1550

OKURA, A., KINOSHITA, T., TANAKA, N. (1971) Formation of fusidic acid-G factor-GDP-ribosome complex and the relationship to the inhibition of GTP hydrolysis. J.Antibiotics 24, 656-660

OKUYAMA, A., MACHIYAMA, N., KINOSHITA, T., TANAKA, N. (1971) Inhibition by kasugamycin of initiation complex formation on 30S ribosomes. Biochem.Biophys.Res.Commun. 43, 196-199

OKUYAMA, A., TANAKA, N. (1972) Differential effects of aminoglycosides on cistron-specific initiation of protein synthesis. Biochem. Biophys.Res.Commun. 49, 951-957

OKUYAMA, A., TANAKA, N. (1973) Studies on the ribosomal binding sites of natural messenger RNA. Biochem.Biophys.Res.Commun. 52, 1463-1469

OKUYAMA, A., TANAKA, N. (1975) The binding of kasugamycin to the *Escherichia coli* ribosomes. J.Antibiotics 28, 903-905

OKUYAMA, A., WATANABE, T., TANAKA, N. (1972) Effects of aminoglycoside antibiotics on initiation of viral RNA-directed protein synthesis. J.Antibiotics 25, 212-218

OKUYAMA, A., YOSHIKAWA, M., TANAKA, N. (1974) Alteration of ribosomal protein S2 in kasugamycin-resistant mutant derived from *Escherichia coli* AB312. Biochem.Biophys.Res.Commun. 60, 1163-1169

OLEINICK, N.L. (1975) In "Antibiotics. III. Mechanism of Action of Antimicrobial and Antitumor Agents" (Corcoran, J.W. and Hahn, F.E., eds.), pp. 396-419, Springer-Verlag: Berlin-Heidelberg-New York

OLEINICK, N.L. (1977) Initiation and elongation of protein synthesis in growing cells: differential inhibition by cycloheximide and emetine. Arch.Biochem.Biophys. 182, 171-180

OLEINICK, N.L., CORCORAN, J.W. (1969) Two types of binding of erythromycin to ribosomes from antibiotic-sensitive and -resistant *Bacillus subtilis* 168. J.Biol.Chem. 244, 727-735

OLEINICK, N.L., CORCORAN, J.W. (1970) Evidence of a limited access of erythromycin A to functional polysomes and its action on bacterial translocation. In "Progress in Antimicrobial and Anticancer Chemotherapy", Vol. 1, pp. 202-208, Tokyo: Tokyo University Press

OLEINICK, N.L., RUSTAD, R.C., KUNCIO, G.S. (1974) Studies on the radiation-induced dissociation and reassociaton of polysomes in *Tetrahymena pyriformis*. Biochim.Biophys.Acta 366, 215-223

OLSNES, S. (1972) Toxic proteins inhibiting protein synthesis. Naturwissenschaften 59, 497-502

OLSNES, S., FERNANDEZ-PUENTES, C., CARRASCO, L., VAZQUEZ, D. (1975b) Ribosome inactivation by the toxin lecting abrin and ricin. I. Kinetics of the enzymic activity of the toxin A-chains. Eur.J. Biochem. 60, 281-288

OLSNES, S., HEIBERG, R., PIHL, A. (1973) Inactivation of eukaryotic ribosomes by the toxic plant proteins abrin and ricin. Mol.Biol. Reports 1, 15-20

OLSNES, S., PAPPENHEIMER, A.M., MEREN, R. (1974) Lectins from *Abrus precatorius* and *Ricinus communis*. II. Hybrid toxins and their interaction with chain-specific antibodies. J.Immunol. 113, 842-847

OLSNES, S., PIHL, A. (1972a) Inhibition of peptide chain elongation. Nature 238, 459-461

OLSNES, S., PIHL, A. (1972b) Ricin a potent inhibitor of protein synthesis. FEBS Letters 20, 327-329

OLSNES, S., PIHL, A. (1972c) Treatment of abrin and ricin with β-mercaptoethanol. Opposite effects of their toxicity in mice and their ability to inhibit protein synthesis in a cell-free system. FEBS Letters 28, 48-50

OLSNES, S., PIHL, A. (1973a) Isolation and properties of abrin: a toxic protein inhibiting protein synthesis. Evidence for different biological functions of its two constituent-peptide chains. Eur.J.Biochem. 35, 179-195

OLSNES, S., PIHL, A. (1973b) Different biological properties of the two constituent peptide chains of ricin, a toxic protein inhibiting protein synthesis. Biochemistry 12, 3121-3126

OLSNES, S., PIHL, A. (1977) Abrin, ricin and their associated agglutinins. In "Receptors and Recognition Series: The Specificity and Action of Animal, Bacterial and Plant Toxins" (Cuatrecasas, E., ed.), pp. 129-173, Chapman and Hall: London

OLSNES, S., REFSNES, K., CHRISTENSEN, T.B., PIHL, A. (1975a) Studies on the structure and properties of the lectins from Abrus precatorius and Ricinus communis. Biochim.Biophys.Acta 405, 1-10

OLSNES, S., REFSNES, K., PIHL, A. (1974) Mechanism of action of the toxic lectins abrin and ricin. Nature 249, 627-631

OLSON, B.H., GOERNER, G.L. (1965) Alpha sarcin, a new antitumor agent. Applied Microbiol. 13, 314-321

OLSON, B.H., JENNINGS, J.C., ROGA, V., JUNEK, A.J., SCHUURMANS, D.S. (1965) Alpha sarcin, a new antitumor agent. Applied Microbiol. 13, 322-326

OLSSON, M., ISAKSSON, L., KURLAND, C.G. (1974) Pleiotropic effects of ribosomal protein S4 studied in Escherichia coli mutants. Mol.Gen. Genetics 135, 191-202

ONO, Y., SCOULTCHI, A., WATERSON, J., LENGYEL, P. (1969) Peptide chain elongation: GTP cleavage catalyzed by factors binding aminoacyl-transfer RNA to the ribosome. Nature 222, 645-658

ONOZAKI, K., HAYATSU, H., UKITA, T. (1975) Inhibition of protein synthesis in mouse myeloma cells by Ricinus communis toxin. Biochim.Biophys.Acta 407, 99-107

ONOZAKI, K., TOMITA, M., SAKURAI, Y., UKITA, T. (1972) The mechanism of the cytotoxicity of Ricinus communis phytoagglutinin toward rat ascites tumor cells. Biochem.Biophys.Res.Commun. 48, 783-788

OPPERMANN, H., KOCH, G. (1976) On the regulation of protein synthesis in vaccinia virus infected cells. J.Gen.Virol. 32, 261-273

O'ROURKE, J.C., GODCHAUX, W. (1975) Fluoride inhibition of the initiation of protein synthesis in the reticulocyte lysate cell-free system. J.Biol.Chem. 250, 3443-3450

OTAKA, E., ITOH, T., OSAWA, S., TANAKA, K., TAMAKI, M. (1971) Peptide analyses of a protein component 50-8 of 50S ribosomal subunit from erythromycin resistant mutants of Escherichia coli and Escherichia freundii. Mol.Gen.Genetics 114, 14-22

OTAKA, T., KAJI, A. (1973) Evidence that fusidic acid inhibits the binding of aminoacyl-tRNA to the donor as well as the acceptor site of the ribosomes. Eur.J.Biochem. 38, 46-53

OTAKA, T., KAJI, A. (1974) Micrococcin: acceptor-site-specific inhibitor of protein synthesis. Eur.J.Biochem. 50, 101-106

OTAKA, T., KAJI, A. (1976) Mode of action of bottromycin A_2. Release of aminoacyl- or peptidyl-tRNA from ribosomes. J.Biol.Chem. 251, 2299-2306

OTAKA, E., TERAOKA, H., TAMAKI, M., TANAKA, K., OSAWA, S. (1970) Ribosomes from erythromycin-resistant mutants of *Escherichia coli* Q13. J.Mol.Biol. 48, 499-510

OUDEGA, B., GRAAF, F.K. de. (1976) Enzymatic properties of cloacin DF13 and kinetics of ribosome inactivation. Biochim.Biophys.Acta 425, 296-304

OUDEGA, B., KLAASEN-BOOR, P., SNEEUWLOPER, G., DeGRAAF, F.K. (1977) Interaction of the complex between cloacin and its immunity protein and of cloacin with the outer and cytoplasmic membranes of sensitive cells. Eur.J.Biochem. 78, 445-453

OWEN, S.P., DIETZ, A., CAMIENER, G.W. (1962) Sparsomycin, a new antitumor antibiotic. I. Discovery and biological properties. Ant.Agents Chemother. 1962, 772-779

OZAKI, M., MIZUSHIMA, S., NOMURA, M. (1969) Identification and functional characterization of the protein controlled by the streptomycin-resistant locus in *E. coli*. Nature 222, 333-339

PAGANO, J.F., WEINSTEIN, M.J., STOUT, H.A., DONOVICK, R. (1955-1956) Thiostrepton, a new antibiotic. In vitro studies. Antibiotics Ann. 554-559

PALMER, E., WILHELM, J.M. (1978) Mistranslation in a eukaryotic organism. Cell 13, 329-334

PAPAS, T.S., SANDHAUS, L., CHIRIGOS, M.A. (1973) Inhibition of DNA polymerase of avian myeloblastosis virus by an alkaloid extract from Narcissus tazetta L. Biochem.Biophys.Res.Commun. 52, 88-92

PAPPENHEIMER, A.M. (1977) Diphtheria toxin. Ann.Rev.Biochem. 46, 69-94

PAPPENHEIMER, A.M., GILL, D.M. (1972) In "Molecular Mechanism of Antibiotic Action" (Muñoz, E., García-Ferrandiz, F. and Vázquez, D., eds.), pp. 134-149. Elsevier, Amsterdam

PAPPENHEIMER, A.M., GILL, D.M. (1973) Diphtheria. Science 182, 353-358

PAPPENHEIMER, A.M., RANDALL, V. (1975) On the alleged high sensitivity of mouse Ehrlich-Letrre ascites tumor cells to diphtheria toxin. Proc.Nat.Acad.Sci. 72, 3149-3152

PARDO, D., ROSSET, R. (1974) Genetic studies of erythromycin resistant mutants of Escherichia coli. Mol.Gen.Genetics 135, 257-268

PARDO, D., ROSSET, R. (1977) Properties of ribosomes from erythromycin resistant mutants of Escherichia coli. Mol.Gen. Genetics 156, 267-271

PASTAN, I.H., JOHNSON, G.S., ANDERSON, W.B. (1975) Role of cyclic nucleotides in growth control. Ann.Rev.Biochem. 44, 491-522

PEDERSEN, F.S.,LUND, E., KJELDGAARD, N.O. (1973) Codon specific, tRNA dependent in vitro synthesis of ppGpp and pppGpp. Nature New Biol. 243, 13-15

PENMAN, S.K., SCHERRER, Y., BECKER, J., DARNELL, E. (1963) Polyribosomes in normal and poliovirus-infected HeLa cells and their relationship to messenger-RNA. Proc. Nat. Acad. Sci. 49, 654-662

PERANI, A., PARISI, B., CARLI, L.de, CIFERRI, O. (1968) Incorporation of amino acids by a cell-free system prepared from human cells cultured in vitro. Biochim. Biophys.Acta 161, 223-231

PESTKA, S. (1968) Studies on the formation of transfer ribonucleic acid-ribosome complexes. V. On the function of a soluble transfer factor in protein synthesis. Proc.Nat.Acad.Sci. 61, 726-733

PESTKA, S. (1969a) Studies on the formation of transfer ribonucleic acid-ribosome complexes.XI. Antibiotic effects on phenylalanyl-oligonucleotide binding to ribosomes. Proc. Nat.Acad.Sci. 64, 709-714

PESTKA, S. (1969b) Translocation, aminoacyl-oligonucleotides, and antibiotic action. Cold Spring Harb.Symp.Quant.Biol. 34, 395-410

PESTKA, S. (1969c) Studies on the formation of transfer ribonucleic acid-ribosome complexes. X. Phenylalanyl-oligonucleotide binding to ribosomes and the mechanism of chloramphenicol action. Biochem.Biophys.Res.Commun. 36, 589-595

PESTKA, S. (1970a) Studies on the formation of transfer ribonucleic acid-ribosome complexes. VIII. Survey of the effect of antibiotics on N-acetyl-phenylalanyl-puromycin formation: possible mechanism of chloramphenicol action. Arch.Biochem. Biophyssics 136, 80-88 (1970)

PESTKA, S. (1970b) Thiostrepton: a ribosomal inhibitor of translocation. Biochem.Biophys.Res.Commun. 40, 667-673

PESTKA, S. (1971) Inhibitors of ribosome functions. In "Annual Review of Microbiology" (Clifton, C.E., ed.), Vol. 25, pp. 487-562, Palo Alto, California, USA

PESTKA, S. (1972a) Peptidyl-puromycin synthesis on polyribosomes from Escherichia coli. Proc.Nat.Acad.Sci. 69, 624-628

PESTKA, S. (1972b) Studies on transfer ribonucleic acid-ribosome complexes.XIX. Effect of antibiotics on peptidyl-puromycin synthesis on polyribosomes from Escherichia coli. J.Biol.Chem. 247, 4669-4678

PESTKA, S. (1974) Antibiotics as probes of ribosome structure: binding of chloramphenicol and erythromycin to polyribosomes; effect of other antibiotics. Ant.Agents Chemother. 5, 255-267

PESTKA, S. (1977) Inhibitors of protein synthesis. In "Molecular Mechanisms of Protein Biosynthesis" (Weissbach, H. and Pestka, S., eds.), pp. 467,553, Academic Press: New York-San Francisco-London

PESTKA, S., BROT, N. (1971) Studies on the formation of transfer ribonucleic acid-ribosome complexes. XV. Effect of antibiotics on steps of bacterial protein synthesis: some new ribosomal inhibitors of translocation. J.Biol.Chem. 246, 7715-7722

PESTKA, S., HINTIKKA, H. (1971) Studies on the formation of ribonucleic acid-ribosome complexes. XVI. Effect of ribosomal translocation inhibitors on polyribosomes. J.Biol.Chem. 246, 7723-7730

PESTKA, S., NAKAGAWA, A., OMURA, S. (1974) Effect of leucomycins and analogues on binding of (^{14}C)erythromycin to Escherichia coli ribosomes. Ant.Agents Chemother. 6, 606-612

PESTKA, S., NIRENBERG, M. (1966) Codeword recognition on 30S ribosomes. Cold Spring Harb.Symp.Quant.Biol. 31, 641-656

PESTKA, S., ROSENFELD, H., HARRIS, R., HINTIKKA, H. (1972) Studies on transfer ribonucleic acid-ribosome complex. XXI. Effect of antibiotics on peptidyl-puromycin synthesis by mammalian polyribosomes. J.Biol.Chem. 247, 6895-6900

PESTKA, S., VINGE, R., DALUGE, S., HARRIS, R. (1973) Effect of puromycin analogues and other agents on peptidyl-puromycin synthesis on polyribosomes. Ant.Agents Chemother. 4, 37-43

PESTKA, S., WEISS, D., VINCE, R., WIENEN, B., STOFFLER, G., SMITH, I. (1976) Thiostrepton-resistant mutants of Bacillus subtilis: localization of resistance to the 50S subunit. Mol.Gen.Genetics 144, 235-241

PETERSON, N.A., RAGHUPATHY, E., McKEAN, C.M. (1971) Inhibition of homogentisic acid of polypeptide synthesis on rat liver and brain ribosomal systems. Biochim.Biophys.Acta 228, 268-281

PIEPERSBERG, W., BOCK, A., YAGUCHI, M., WITTMANN, H.G. (1975) Genetic position and amino acid replacements of several mutations in ribosomal protein S5 from Escherichia coli. Mol.Gen.Genetics 143, 43-52

PINPHANICHAKARN, P., KRAMER, G., HARDESTY, B. (1976) Partial reaction of peptide initiation inhibited by the reticulocyte hemin-controlled repressor. Biochem.Biophys.Res.Commun. 73, 625-631

PINPHANICHAKARN, P., KRAMER, G., HARDESTY, B. (1977) Partial purification and characterization of a translational inhibitor from Friend leukemia cells. J.Biol.Chem. 252, 2106-2112

PIRALI, G., LANCINI, G.C., PARISI, B., SALA, F. (1972) Interaction of sporangiomycin with the bacterial ribosome. J.Antibiotics 25, 561-568

PIRALI, G., SOMMA, S., LANCINI, G.C., SALA, F. (1974) Inhibition of peptide chain initiation in Escherichia coli by thermorubin. Biochim.Biophys.Acta 366, 310-318

PITILLO, R.F., FISHER, M.W., McALPINE, R.J., THOMPSON, P.E., EHRLICH, J. (1959) Actinobolin, a new broad-spectrum antibiotic. Origin and biological evaluation. Antibiotics Annual 1958-1959, 497-504

POCHE, H., JUNGHAHN, I., GEISSLER, E., BIELKA, H. (1975) Cycloheximide resistance in chinese hamster cells. III. Characterization of cell-free protein synthesis by polysomes. Mol.Gen.Genetics 138, 173-177

PONG, S.S., NUSS, D.L., KOCH, G. (1975) Inhibition of initiation of protein synthesis in mammalian tissue culture cells by L-1-tosylamido-2-phenylethyl-chloromethyl-ketone. J.Biol.Chem. 250, 240-245

PONGRATZ, M., KLINGMULLER, W. (1973) Role of ribosomes in cycloheximide resistance of Neurospora mutants. Mol.Gen.Genetics 124, 359-363

PONGS, O., BALD, R., ERDMANN, V.A. (1973) Identification of chloramphenicol binding protein in Escherichia coli ribosomes by affinity labelling. Proc.Nat.Acad.Sci. 70, 2229-2233

PONGS, O., BALD, R., WAGNER, T., ERDMANN, V.A. (1973) Irreversible binding of N-iodoacetylpuromycin to E. coli ribosomes. FEBS Letters 35, 137-140

PONGS, O., ERDMANN, V.A. (1973) Affinity labeling of E. coli ribosomes with a streptomycin-analogue. FEBS Letters 37, 47-50

PONGS, O., MESSER, W. (1976) The chloramphenicol receptor site in Escherichia coli in vivo affinity labeling with monoiodo-amphenicol. J.Mol.Biol. 101, 171-184

POPELAK, A., HAACK, E., SPINGLER, H. (1966) Uber ein neues alkaloid aus Alangium lamarckii Thw. Tetrahedron Letters 1081-1085

PORTER, J.N., HEWITT, R.I., HESSELTINE, C.W., KRUPKA, G., LOWERY, J.A., WALLACE, W.S., BOHONOS, N., WILLIAMS, J.H. (1952) A new antibiotic having trypanocidal properties. Antibiotics Chemother. 2, 409-415

POWELL, R.G., WEISLEDER, D., SMITH, C.R. (1972) Antitumor alkaloids from Cephalotaxus harringtonia: structure and activity. J.Pharm. Sciences 61, 1227-1230

POZDNYAKOV, V.A., MITIN, Y.V., KUKHANOVA, M.K., NIKOLAEVA, L.V., KRAYEVSKY, A.A., GOTTIKH, B.P. (1972) On the mechanism of peptide bond synthesis in the ribosome 3'-O-phenylalanyl-2'-O-methyladenosine as a peptide acceptor. FEBS Letters 24, 177-180

PRANGE, T., DACRUIX, A., PASCARD, C., LUNEL, J. (1977) Structure of nosiheptide, a polythiazole-containing antibiotic. Nature 265, 189-190

PRYOR, J., BERTHET, J. (1960) The action of adenosine 3', 5'-monophosphate on the incorporation of leucine into liver proteins. Biochim.Biophys.Acta 43, 556-557

RAJALAKSHMI, S., LIANG, H., SARMA, D.S.R., KISILEVSKY, R., FARBER, E. (1971) Cycloheximide, an inhibitor of peptide chain termination or release in liver in vivo and in vitro. Biochem.Biophys.Res. Commun. 42, 259-265

RAMIREZ, G., LEVITAN, I.B., MUSHYNSKI, W.E. (1972) Amino acid incorporation into nerve ending structures in vitro. FEBS Letters 21, 17-20

RANKINE, A.D., LEADER, D.P. (1975) The phosphorylation of ascites cell ribosomes in vivo: identification of a phosphorylated protein of the small ribosomal subunit by two-dimensional gel electrophoresis. FEBS Letters 52, 284-287

RAO, S.S., GROLLMAN, A.P. (1967) Cycloheximide resistance in yeast: a property of the 60S ribosomal subunit. Biochem.Biophys.Res. Commun. 29, 696-704

RATNER, L., SEN, G.C., BROWN, G.E., LEBLEU, B., KAWAKITA, M., CABRER, B., SLATTERY, E., LENGYEL, P. (1977) Interferon, double-stranded RNA and RNA degradation. Eur.J.Biochem. 79, 565-577

RAVEL, J.M., SHOREY, R.L., FROEHNER, S., SHIVE, W. (1968) A study of the enzymic transfer of aminoacyl-tRNA to Escherichia coli ribosomes. Arch.Biochem.Biophys. 125, 514-526

RAVEL, J.M., SHOREY, R.L., SHIVE, W. (1970) Relationship between peptidyl transferase activity and interaction of ribosomes with phenylalanyl transfer ribonucleic acid-guanosine 5'-triphosphate-Tlu complex. Biochemistry 9, 5028-5033

REEVES, P. (1972) The Bacteriocins. Molecular Biology Biochemistry and Biophysics (A. Kleinzeller, G.F. Springer and H.G. Wittmann, eds.), Vol. 11, 138 pp., Springer-Verlag: Berlin-Heidelberg-New York

REFSNES, K., HAYLETT, T., SANDVIG, K., OLSNES, S. (1977) Modeccin, a plant toxin inhibiting protein synthesis. Biochem.Biophys.Res. Commun. 79, 1176-1183

REFSNES, K., OLSNES, S., PIHL, A. (1974) On the toxic proteins abrin and ricin. Studies of their binding to and entry into Ehrlich ascites cells. J.Biol.Chem. 249, 3557-3562

REICHMAN, M., PENMAN, S. (1973) The mechanism of inhibition of protein synthesis by 5-azacytidine in HeLa cells. Biochim.Biophys.Acta 324, 282-289

REISNER, A.H., BUCHOLTZ, C. (1977) The in vivo effect of dimethil sulphoxide (DMSO) on protein synthesis and the polyribosome prodile in Paramecium. J.Cell.Physiol. 90, 169-177

REMY, P., ENGEL, M.L., DIRHEIMER, G., EBEL, J.P., REVEL, M. (1970) Analogues of nucleoside polyphosphates. II. Guanosine-5'-phosphohypophosphate: preparation and effects on protein synthesis in vitro. J.Mol.Biol. 48, 173-176

RENDI, R., OCHOA, S. (1961) Effect of chloramphenicol on protein synthesis in cell-free preparations of Escherichia coli. J.Biol. chem. 237, 3711-3713

REUSSER, F. (1969) Mode of action of berninamycin. An inhibitor of protein biosynthesis. Biochemistry 8, 3303-3308

REUSSER, F. (1973) Rubradirin, an inhibitor of ribosomal polypeptide biosynthesis. Biochemistry 12, 1136-1142

REUSSER, F. (1975) Effect of lincomycin and clindamycin on peptide chain initiation. Ant.Agents Chemother. 7, 32-37

REYES, R., VAZQUEZ, D., BALLESTA, J.P.G. (1976a) Activities of nucleoprotein particles derived from rat liver ribosome. Biochim. Biophys.Acta 435, 317-332

REYES, R., VAZQUEZ, D., BALLESTA, J.P.G. (1976b) Structure and function of rat liver ribosomes. Modification by methoxy-5-nitrotropone treatment. Eur.J.Biochem. 67, 267-274

REYES, R., VAZQUEZ, D., BALLESTA, J.P.G. (1977) Peptidyl transferase center of rat liver ribosome cores. Eur.J.Biochem. 73, 25-31

RICHMAN, N., BODLEY, J.W. (1972) Ribosomes cannot interact simultaneously with elongation factors EF Tu and EF G. Proc.Nat. Acad.Sci. 69, 686-689

RICHMAN, N., BODLEY, J.W. (1973) Irreversible inhibition of the interaction between elongation factor Tu and phenylalanyl transfer ribonucleic acid by L-1-tosylamido-2-phenetyl-chloromethyl-ketone. J.Biol.Chem. 248, 381-383

RICHTER, D. (1972) Inability of E. coli ribosomes to interact simultaneously with the bacterial elongation factors EF Tu and EF G. Biochem.Biophys.Res.Commun. 46, 1850-1856

RINEHART, K.L., MICHENS, M., ARGOUNDELIS, A.D., CHILTON, W.S., CARTER, H.E., GEORGIADIS, M.P., SCHAFFNER, C.P., SCHILLINGS, R.T. (1962) Chemistry of the neomycins: X: Neomycins B and C. J.Amer. Chem.Soc. 84, 3218-3220

ROBERTS, W.K., CLEMENS, M.J., KERR, I.M. (1976) Interferon-induced inhibition of protein synthesis in L-cell extracts: an ATP-dependent step in the activation of an inhibitor by double-stranded RNA. Proc.Nat.Acad.Sci. 73, 3136-3140

ROBERTSON, H.D., MATTHEWS, H.B. (1973) Double-stranded RNA as an inhibitor of protein synthesis and as a substrate for a nuclease in extracts of Krebs II ascites cells. Proc.Nat.Acad.Sci. 70, 225-229

ROBERTUS, J.D., MONZING, A.F., IRVIN, J.D. (1977) Preliminary X-ray diffraction studies on an antiviral protein. Biochem.Biophys.Res. Commun. 74, 775-779

ROBINSON, E.A., MAXWELL, E.S. (1972) Chemical properties of elongation factor 2. Amino acid composition, NH_2-terminal residue, and sulphydryl reactivity. J.Biol.Chem. 21, 7023-7028

RODRIGUEZ-LOPEZ, M., VAZQUEZ, D. (1968) Comparative studies on cytoplasmic ribosomes from algae. Life Sci. 7, 327-336

ROMAN, R., BROOKER, J.D., SEAL, S.N., MARCUS, A. (1976) Inhibition of the transition of a 40S ribosome-Met-$tRNA_i^{Met}$ complex to an 80S ribosome-Met-$tRNA_i^{Met}$ complex by 7-methylguanosine-5'-phosphate. Nature 260, 359-360

ROSEN, S.W., HUGHES, R.C. (1977) Effects of neuraminidase on lectin binding by wild-type and ricin-resistant strains of hamster fibroblasts. Biochemistry 16, 4908-4915

ROSETT, T., SANKHALA, R.H., STICKINGS, C.E., TAYLOR, M.E.U., THOMAS, R. (1957) Studies in the biochemistry of microorganisms. 103. Metabolites of Alternaria tenuis Auct.: culture filtrate products. Biochem.J. 67, 390-400

RYCHLIK, I. (1966) Release of lysine peptides by puromycin from polylysyl-transfer RNA in presence of ribosomes. Biochim.Biophys. Acta 114, 425-427

RYCHLIK, I., CERNA, J. (1972) Effect of macrolide antibiotics on ribosomal peptidyl transferase. In "Proc. 7th International Congress of Chemotherapy, Vol. I" (Semonsky, M., Hejzlar, M. and Masak, S., eds.) pp. 793-799, Urban and Schwarzenberg: Munich-Berlin-Viena

RYCHLIK, I., CERNA, J., CHLADEK, S., ZEMLICKA, J., HALADOVA, Z. (1969) Substrate specificity of ribosomal peptidyl transferase: 2'(3')-O-aminoacyl nucleosides as acceptors of the peptide chain on the amino acid site. J.Mol.Biol. 43, 13-24

RYCHLIK, I., CERNA, J., ZEMLICKA, J., CHLADEK, S. (1970) Substrate specificity of ribosomal peptidyl transferase. II. 2'(3')-O-aminoacyl nucleosides as acceptors of the peptide chain in the fragment reaction. Biochim.Biophys.Acta 204, 203-209

SABORIO, J.L., PONG, S.S., KOCH, G. (1974) Selective and reversible inhibition of initiation of protein synthesis in mammalian cells. J.Mol.Biol. 85, 195-211

SAKAI, F., TAKEBE, I. (1970) RNA and protein synthesis in protoplasts isolated from tobacco leaves. Biochim.Biophys.Acta 224, 531-540

SAKAKIBARA, H., NAGANAWA, H., OHNO, M., MAEDA, K., UMEZAWA, H. (1974) The structure of althiomycin. J.Antibiotics 27, 897-899.

SALA, F., KUNTZEL, H. (1970) Peptide chain initiation in homologous and heterologous systems from mitochondria and bacteria. Eur.J. Biochem. 15, 280-286

SAN-MILLAN, M.J., VAZQUEZ, D., MODOLELL, J. (1975) The interaction of fusidic acid with peptidyl-transfer-ribonucleic acid·ribosome complexes. Eur.J.Biochem. 57, 431-440

SANCHEZ, L., VAZQUEZ, D., JIMENEZ, A. (1977) Genetics and biochemistry of cryptopleurine resistance in the yeast *Saccharomyces cerevisiae*. Mol.Gen.Genetics 156, 319-326

SANDER, G. (1977a) Mechanism of action of colicin E3. Effect on ribosomal elongation-factor-dependent reactions. Eur.J.Biochem. 75, 523-531

SANDER, G. (1977b) Role of components of the 50S ribosomal subunit in colicin E3 activity. FEBS Letters 83, 293-296

SANDVIG, K., Olsnes, S., PIHL, A. (1978) Binding, uptake and degradation of the toxic proteins abrin and ricin by toxin-resistant cell variants. Eur.J.Biochem. 82, 13-23

SARKAR, P., STRINGER, E.A., MAITRA, U. (1974) Thiostrepton inhibition of initiation factor 1 activity in polypeptide chain initiation in *Escherichia coli*. Proc.Nat.Acad.Sci. 71, 4986-4990

SCHATZ, A., BUGIE, E., WAKSMAN, S.A. (1944) Streptomycin, a substance exhibiting antibiotic activity against gram-positive and gram-negative bacteria. Proc.Soc.Expt.Biol.Med. 55, 66-69

SCHINDLER, D. (1974) Two classes of inhibitors of peptidyl transferase activity in eukaryotes. Nature 249, 38-41

SCHINDLER, D.G., DAVIES, J.E. (1977) Specific cleavage of ribosomal RNA caused by alpha sarcin. Nucleic Acids Res. 4, 1097-1110

SCHINDLER, D., GRANT, P., DAVIES, J. (1974) Trichodermin resistance-mutation affecting eukaryotic ribosomes. Nature 248, 535-536

SCHMITZ, H., FARDIG, O.B., O'HERRON, F.A., ROUSHE, M.A., HOOPER, I.R. (1958) Kanamycin III. Kanamycin B. J.Amer.Chem.Soc. 80, 2911-2912

SCHNEIDER, J.A., RAEBURN, S., MAXWELL, E.S. (1968) Translocase activity in the aminoacyl transferase II fraction from rat liver. Biochem.Biophys.Res.Commun. 33, 177-181

SCHREINER, G., NIERHAUS, K.H. (1973) Protein involved in the binding of dihydrostreptomycin to ribosomes of Escherichia coli. J.Mol. Biol. 81, 71-82

SCHRIER, P.I., MOLLER, W. (1975) The involvement of 50S ribosomal protein L11 in the EF-G dependent GTP hydrolysis of E. coli ribosomes. FEBS Letters 54, 130-134

SCOLNICK, E., TOMPKINS, R., CASKEY, T., NIRENBERG, M. (1968) Release factors differing in specificity for termination codons. Proc. Nat.Acad.Sci. 61, 768-774

SCORNIK, O.A. (1974) In vivo rates of deaggregation of polyribosomes in normal and regenerating liver after the injection of pactamycin. Biochim.Biophys.Acta 374, 76-81

SEAL, S.N., MARCUS, A. (1972) Reactivity of ribosomally bound methionyl-tRNA with puromycin and the locus of pactamycin inhibition of chain initiation. Biochem.Biophys.Res.Commun. 46, 1895-1902

SEDLACEK, J., JONAK, J., RYCHLIK, I. (1971) Inactivation of protein-synthesizing T-factor by N-tosyl-L-phenylalanyl chloromethane. Biochim.Biophys.Acta 254, 478-480

SELLERS, A., BLOXHAM, D.P., MUNDAY, K.A., AKHTAR, M. (1974) Antianabolic effects of adenosine 3':5'-cyclic monophosphate. Inhibition of protein synthesis. Biochem.J. 138, 335-340

SEN, G.C., SHAILA, S., LEBLEU, B., BROWN, G.E., DESROISIERS, R.C., LENGYEL, P. (1977) Impairment of reovirus mRNA methylation in extracts of interferon-treated Ehrlich ascites tumor cells: further characteristics of the phenomenon. J.Virol. 21, 69-83

SENIOR, B.W., HOLLAND, I.B. (1971) Effect of colicin E3 upon the 30S ribosomal subunit of Escherichia coli. Proc.Nat.Acad.Sci. 68, 959-963

SHAFRITZ, D.A., GOODWIN, F., WEISSBACH, H. (1969) Inhibition by aminochromes of in vitro polypeptide synthesis in Escherichia coli. Arch.Biochem.Biophys. 134, 478-485

SHAFRITZ, D.A., WEINSTEIN, J.A., SAFER, B., MERRICK, W.C., WEBER, L.A., HICKEY, E.D., BAGLIONI, C. (1976) Evidence for role of $m^7G^{5'}$-phosphate group in recognition of eukaryotic mRNA by initiation factor IF-M_3. Nature 261, 291-294

SHARMA, O.K., HRUBY, D.E., BEEZLEY, D.N. (1976) Inhibition of ovalbumin mRNA translation by 7-methylguanosine-5'-phosphate. Biochem.Biophys.Res.Commun. 72, 1392-1398

SHERMAN, M.I., SIMPSON, M.U. (1969) The role of ribosomal conformation in protein biosynthesis: the streptomycin-ribosome interaction. Proc.Nat.Acad.Sci. 64, 1388-1395

SHERTON, C.C., WOOL, I.G. (1972) Determination of the number of proteins in liver ribosomes and ribosomal subunits by two-dimensional polyacrylamide gel electrophoresis. J.Biol.Chem. 247, 4460-4467

SHIER, W.T., KENNETH,L., RINEHART, K.L., GOTTLIEB, D. (1969) Preparation of four new antibiotics from a mutant of streptomyces fradiae. Proc.Nat.Acad.Sci. 63, 198-204

SHIER, W.T., RINEHART, K.L., GOTTLIEB, D. (1970) Preparation of two new aminoglycoside antibiotics. J.Antibiotics 23, 51-53

SHIGEURA, H.T., GORDON, C.N. (1963) The biological activity of tenuazonic acid. Biochemistry 2, 1132-1137

SHIH, D.S., DASGUPTA, R., KAESBERG, P. (1976) 7-Methyl-guanosine and efficiency of RNA translation. J.Virol. 19, 637-642

SHIMIZU, M., SAITO, T., MITSUHASHI, S. (1970) Macrolide resistance in Staphylococcus aures. Decrease of spiramycin-binding to 50S ribosomal subunit in macrolide resistant strains of Staphylococci. J.Antibiotics 23, 467-468

SHINE, J., DALGARNO, L. (1975) Determinant of ustron specificity in bacterial ribosomes. Nature 254, 34-38

SHINOZAWA, T., YAHARA, I., IMAHORI, K. (1968) Interaction of polyvinyl sulphate with ribosomes. J.Mol.Biol. 36, 305-319

SHOJI, J., HINOO, H., WAKISAKA, Y., KOIZUMI, K., MAYAMA, M., MATSUURA, S., MATSUMOTO, K. (1976) Isolation of three new antibiotics, thiocillins, I, II and III, related to micrococcin P. J.Antibiotics 29, 366-374

SHOMURA, T., EZAKI, N., TSURNOKA, T., NIWA, T., AKITA, E., NIIDA, T. (1970) Studies on antibiotics SF-733, a new antibiotic. I. Taxonomy, isolation and characterization. J.Antibiotics 23, 155-161

SIDIKARO, J., NOMURA, M. (1973) Colicin E3-induced in vitro inactivation of ribosomes from colicin-insensitive bacterial species. FEBS Letters 29, 15-19

SIEGEL, M.R., SISLER, H.D. (1963) Inhibition of protein synthesis in vitro by cycloheximide. Nature 200, 675-676

SIEGEL, M.R., SISLER, H.D. (1965) Site of action of cycloheximide in cells of Saccharomyces pastorianus. III. Further studies on the mechanism of action and the mechanism of resistance in saccharomyces species. Biochim.Biophys.Acta 103, 558-567

SIEGELMAN, F., APIRION, D. (1971) Aurintricarboxylic acid a preferential inhibitor of initiation of protein synthesis. J.Bacteriol. 105, 902-907

SIERRA, J.M., HARO, C., DATTA, A., OCHOA, S. (1977) Translational control by protein kinases in Artemia salina and wheat germ. Proc.Nat.Acad.Sci. 74, 4356-4359

SILVERMAN, R.H., ATHERLY, A.G. (1978) Unusual effects of 5a, 6-anhidrotetracycline and other tetracyclines: inhibition of guanosine 5'-diphosphate 3'-diphosphate metabolism, RNA accumulation and other growth related processes in Escherichia coli. Biochim.Biophys.Acta (in press)

SINOHARA, H., SKY-PECK, H.H. (1965) Effect of gougerotin on the protein synthesis in the mouse liver. Biochem.Biophys.Res. Commun. 18, 98-102

SISLER, H.D., SIEGEL, M.R. (1967) Cycloheximide and other glutarimide antibiotics. In "Antibiotics. Vol. I" (Gottlieb, D. and Shaw, P.D., eds.), pp. 283-307, Springer-Verlag: Berlin-Heidelberg-New York

SKOGERSON, L., McLAUGHLIN, C., WAKATAMA, E. (1973) Modification of ribosomes in cryptopleurine-resistant mutants of yeast. J. Bacteriol. 116, 818-822

SKORVE, J., ABRAHAM, A.K., OLSNES, S., PIHL, A. (1977) Effect of abrin on peptide chain initiation. Eur.J.Biochem. 79, 559-564

SKOULTCHI, A., ONO, Y., WATERSON, J., LENGYEL, P. (1970) Peptide chain elongation: indications for the binding of an amino acid polymerization factor, guanosine 5'-triphosphate-aminoacyl transfer ribonucleic acid complex to the messenger ribosome complex. Biochemistry 9, 508-514

SLOBIN, L.I. (1976) The inhibition of elongation factor 1 activity by brain. Biochem.Biophys.Res.Commun. 73, 539-547

SMITH, K.E., CANNON, M., CUNDLIFFE, E. (1975) Inhibition at the initiation level of eukaryotic protein synthesis by T-2 toxin. FEBS Letters 50, 8-12

SMITH, I.S., WEISS, D., PESTKA, S. (1976) A micrococcin-resistant mutant of Bacillus subtilis: localization of resistance to the 50S subunit. Mol.Gen.Genetics 144, 231-233

SMITHERS, D., BENNET, L.L., STRUCK, R.F. (1969) Inhibition of protein synthesis in mammalian cells by actinobolin. Mol.Pharmacol. 5, 433-445

SO, A.G., BODLEY, J.W., DAVIE, E.W. (1964) The influence of environment on the specificity of polynucleotide-dependent amino acid incorporation into polypeptides. Biochemistry 3, 1977-1982

SO, A.G., DAVIE, E.W. (1964) The effect of organic solvents on protein biosynthesis and their influence on the amino acid code. Biochemistry 3, 1165-1169

SOBIN, B.A., TANNER, F.W. (1954) Anisomycin, a new antiprotozoan antibiotics. J.Amer.Chem.Soc. 76, 4053

SONENBERG, N., WILCKEK, M., ZAMIR, A. (1973) Mapping of Escherichia coli ribosomal components involved in peptidyl transferase activity. Proc.Nat.Acad.Sci. 70, 1423-1426

SOPORI, M.L., LENGYEL, P. (1972) Components of the 50S ribosomal subunit involved in GTP cleavage. Biochem.Biophys.Res.Commun. 46, 238-244

SPARLING, P.F. (1970) Kasugamycin resistance: 30S ribosomal mutation with an unusual location on the *Escherichia coli* chromosome. Science 167, 56-58

SPERTI, S., MONTANARO, L., MATTIOLI, A., STIRPE, F. (1973) Inhibition by ricin of protein synthesis in vitro: 60S ribosomal subunit as the target of the toxin. Biochem.J. 136, 813-815

SPERTI, S., MONTANARO, L., MATTIOLI, A., TESTONI, G. (1975) Relationship between elongation factor 1 and elongation factor 2 dependent guanosine triphosphatase activities of ribosomes. Inhibition of both activities by ricin. Biochem.J. 148, 447-451

SPERTI, S., MONTANARO, L., MATTIOLI, A., TESTONI, G., STIRPE, F. (1976) Inhibition of protein synthesis in vitro by crotins and ricin. Effect on the steps of peptide elongation. Biochem.J. 156, 7-13

SPEYER, J.F., LENGYEL, P., BASILIO, C. (1962) Ribosomal localization of streptomycin sensitivity. Proc.Nat.Acad.Sci. 48, 684-686

SPEYER, J.F., LENGYEL, P., BASILIO, C., WAHBA, A.J., GARDNER, R.S., OCHOA, S. (1963) Synthetic polynucleotides and the amino acid code. Cold Spring Harb.Symp.Quant.Biol. 28, 559-567

SPIRIN, A.S. (1972) In "Molecular Mechanism of Antibiotic Action on Protein Biosynthesis and Membranes" (Muñoz, E., García-Ferrandiz, F. and Vázquez, D., eds.), pp. 11-27, Elsevier: Amsterdam

SPOTTS, C.R., STANIER, R.Y. (1961) The mechanism of streptomycin action on bacteria: a unitary hypothesis. Nature 192, 633-637

STAFFORD, M.E., McLAUGHLIN, C.S. (1973) Trichodermin, a possible inhibitor of the termination process of protein synthesis. J.Cell. Physiol. 82, 121-128

STAHL, J., DRESSLER, K., BIELKA, H. (1974) Studies on proteins of animal ribosomes. Affinity labeling of rat liver ribosomes by N-bromoacetylpuromycin. FEBS Letters 47, 167-170

STANLEY, W.M., DETTMAN, G.L. (1973) The ternary complex of initiator factor IF 1, Met-$tRNA_f^{Met}$ and GTP. Biochim.Biophys.Acta 299, 142-147

STANNERS, C.P. (1966) The effect of cycloheximide on polyribosomes from hamster cells. Biochem.Biophys.Res.Commun. 24, 758-764

STEPHENS, C.R., CONOVER, L.H., PASTERNAK, R., HOCHSTEIN, F.A., MORELAND, W.T., REGNA, P.P., PILGRIM,F.J., BRUNNINGS, K.J., WOODWARD, R.B. (1954) The structure of aureomycin. J.Amer.Chem. Soc. 76, 3568-3572

STEVENS, C.L., MEMEC, J., RANSFORD, J.H. (1972) Total synthesis of the amino sugar nucleoside antibiotic plicacentin. J.Amer.Chem.Soc. 94, 3280-3281

STEWART, M.L., GOLDBERG, I.H. (1973) Pactamycin binding to Escherichia coli ribosomes. Interference by formation of the protein synthesizing complex with f_2 viral RNA. Biochim.Biophys. Acta 294, 123-137

STEWART, M.L., GROLLMAN, A.P., HUANG, M.T. (1971) Aurintricarboxylic acid: inhibitor of initiation of protein synthesis. Proc.Nat. Acad.Sci. 68, 97-101

STEWART-BLAIR, M.L., YANOWITZ, I.S., GOLDBERG, I.H. (1971) Inhibition of synthesis of new globin chains in reticulocyte lysates by pactamycin. Biochemistry 10, 4198-4206

STIRPE, F.,GASPERI-CAMPANI, A., BARBIERI, L., LORENZONI, E., MONTANARO, L., SPERTI, S., BONETTI, E. (1978) Inhibition of protein synthesis by modeccin, the toxin of Modecca digitata. FEBS Letters 85, 65-67

STIRPE, F., PESSION-BRIZZI, A., LORENZONI, E., STROCCHI, P., MONTANARO, L., SPERTI, S. (1976) Studies on the proteins from the seeds of Croton tiglium and of Jatropha curcas. Biochem.J. 156, 1-6

STOUTHAMER, A.H., TIEZE, G.A. (1966) Bacteriocin production by members of the genus klebsiella. Antonie van Leeuwenhoek 32, 171-182

STRELTSOV, S.A., KUKHANOVA, M.K., KRAYEVSKY, A.A., BELJAVSKAJA, I.V., VICTOROVA, L.S., GURSKY, G.V., TREBOGANOV, A.D., GOTT, Kh, B.P. (1974) Binding of oxytetracycline to E. coli ribosomes. Mol.Biol. Reports 1, 391-396

SU, T.L. (1948) Micrococcin, an antibacterial substance formed by a strain of micrococcus. Brit.J.Exp.Pathol. 29, 473.

SUAREZ, G., NATHANS, D. (1965) Inhibition of aminoacyl-tRNA binding to ribosomes by tetracycline. Biochem.Biophys.Res.Commun. 18, 743-750

SUHADOLNIK, R.J. (1970) Nucleoside Antibiotics. Wiley Interscience: New York-London

SUITS, J.P., IRVIN, J.D. (1976) The nature of the differential inhibition of polyphenylalanine synthesis in extracts from Artemia salina and rabbit reticulocytes by the Phytolacca americana protein. Mol.Biol.Reports 2, 363-367

SUNDARAM, P.V. (1978) In "Enzyme Engineering" (Wingard, L., Jr., ed.), Vol. 4, (in press), Plenum Press: New York-London.

SUNG, S. (1972) Metabolic Inhibitors (Hochster, R.M., Kates, M. and Quastel, J.H., eds.), Vol. III, pp. 175-204, Academic Press: New York-London

SUZUKA, I., KAJI, H., KAJI, A. (1966) Binding of specific sRNA to 30S ribosomal subunits: effect of 50S ribosomal subunits. Proc. Nat.Acad.Sci. 55, 1483-1490

SUZUKI, H. (1977) Effect of $m^7G^{5'}ppp^{5'}Nm$ on the rabbit globin synthesis. FEBS Letters 79, 11-14

SUZUKI, H., GOLDBERG, I.H. (1974) Reversal of pactamycin inhibition of methionyl-puromycin synthesis and 80S initiation complex formation by a ribosomal joining factor. Proc.Nat.Acad.Sci. 71, 4259-4263

SWART, E.A., LECHEVALIER, H.A., WAKSMAN, S.A. (1951) The identity of the neomycin complex as measured by countercurrent distribution and microbial analyses. J.Chem.Soc. 73, 3253-3255

SZER, W., HERMOSO, J.M., LEFFLER, S. (1975) Ribosomal protein S1 and polypeptide chain initiation in bacteria. (30S subunit subspecies/gel electrophoresis/initiation complex formation). Proc.Nat.Acad.Sci. 72, 2325-2329

SZER, W., KURYLO-BOROWSKA, Z. (1972) Interactions of edeine with bacterial ribosomal subunits. Selective inhibition of aminoacyl-tRNA binding sites. Biochim.Biophys.Acta 259, 357-368

SZER, W., OCHOA, S. (1964) Complexing ability and coding properties of synthetic polynucleotides. J.Mol.Biol. 8, 823-834

TAI, P.C. (1975) Impaired initiation complex formation on ribosomes treated with colicin E3. Biochem.Biophys.Res.Commun. 67, 1466-1472

TAI, P.C., DAVIS, B.D. (1974) Activity of colicin E3-treated ribosomes in initiation and in chain elongation. Proc.Nat.Acad.Sci. 71, 1021-1025

TAI, P.C., WALLACE, B.J., DAVIS, B.D. (1974) Selective action of erythromycin on initiating ribosomes. Biochemistry 13, 4653-4659

TAI, P.C., WALLACE, B.J., DAVIS, B.D. (1978) Streptomycin causes misreading of natural messenger by interacting with ribosomes after initiation. Proc.Nat.Acad.Sci. 75, 275-279

TAKAHASHI, Y., NAGANAWA, H., TAKITA, T., UMEZAWA, H., NAKAMURA, S. (1976) The revised structure of bottromycin A_2. J.Antibiotics 29, 1120-1123

TAKEUCHI, S., HIRAYAMA, K., UEDA, K., SAKAI, H., YONEHARA, H. (1958) Blasticidin S, a new antibiotic. J.Antibiotics 11, 1-5

TANAKA, K., TAMAKI, M., ITOH, T., OTAKA, E., OSAWA, S. (1971) Ribosomes from spiramycin resistant mutants of Escherichia coli Q13. Mol.Gen.Genetics 114, 23-30

TANAKA, K., TERAOKA, H., TAMAKI, M. (1971) Peptidyl puromycin synthesis: effect of several antibiotics which act on 50S ribosomal subunits. FEBS Letters 13, 65-67

TANAKA, K., WATANABE, S., TERAOKA, H., TAMAKI, M. (1970) Effect of siomycin on protein synthesizing activity of Escherichia coli ribosomes. Biochem.Biophys.Res.Commun. 39, 1189-1193

TANAKA, N. (1975) In "Antibiotics. III. Mechanism of Action of Antimicrobial and Antitumor Agents" (Corcoran, J.W. and Hahn, F.E., eds.), pp. 487-497, Springer-Verlag: Berlin-Heidelberg-New York

TANAKA, N., IGUSA, S. (1968) Effects of viomycin and polymixin B on protein synthesis in vivo. J.Antibiotics 21, 239-240

TANAKA, N., KINOSHITA, T., MASUKAWA, H. (1968) Mechanism of protein synthesis inhibition by fusidic acid and related steroidal antibiotics. Biochem.Biophys.Res.Commun. 30, 278-283

TANAKA, N., LIN, Y.C., OKUYAMA, A. (1971) Studies on translocation of F-Met-tRNA and peptidyl-tRNA with antibiotics. Biochem. Biophys.Res.Commun. 44, 477-483

TANAKA, N., MASUKAWA, H., UMEZAWA, H. (1967) Structural basis of kanamycin for miscoding activity. Biochem.Biophys.Res.Commun. 26, 544-549

TANAKA, N., NISHIMURA, T., KINOSHITA, T., UMEZAWA, H. (1969) The effect of fusidic acid on protein synthesis in a mammalian system. J.Antibiotics 22, 181-182

TANAKA, N., SASHIKATA, K., UMEZAWA, H. (1967) Antibiotic-sensitivity of ribosomes from kanamycin-resistant E. coli. J.Antibiotics 20, 115-119

TANAKA, N., YAMAGUCHI, H., UMEZAWA, H. (1966) Mechanism of kasugamycin action on polypeptide synthesis. J.Biochem. 60, 429-434

TANAKA, S., OTAKA, T., KAJI, A. (1973) Further studies on mechanism of erythromycin action. Biochim.Biophys.Acta 331, 128-140

TANAKA, T., ENDO, T., SHIMAZU, A., YOSHIDA, R., SUZUKI, Y., OTAKE, N., YONEHARA, H. (1970) A new antibiotic, multhiomycin. J.Antibiotics 23, 231-237

TANAKA, T., SAKAGUCHI, K., YONEHARA, H. (1970) On the mode of action of multhiomycin. I. Effects of multhiomycin on macromolecular synthesis. J.Antibiotics 23, 401-407

TANAKA, T., SAKAGUCHI, K., YONEHARA, H. (1971) Inhibition by multhiomycin of T factor- and GTP-dependent binding of phenylalanyl-tRNA to ribosomes and hydrolysis associated with it. J.Biochem. 69, 1127-1130

TANNER, F.W., SOBIN, B.A., GARDOCKI, J.F. (1955) Some chemical and biological properties of anisomycin. Antibiotics Annual 1954-1955, 809

TATE, W.P., CASKEY, C.T. (1973) Peptidyl transferase inhibition by trichodermin. J.Biol.Chem. 248, 7970-7972

TATE, W.P., CASKEY, C.T. (1974) The mechanism of peptide chain termination. Mol.Cell.Biochem. 5, 115-126

TAUBMAN, S.B., YOUNG, F.E., CORCORAN, J.W. (1963) Antibiotic glycosides. IV. Studies on the mechanism of erythromycin resistance in Bacillus subtilis. Proc.Nat.Acad.Sci. 50, 955-962

TERAOKA, H. (1970) A reversible change in the ability of Escherichia coli ribosomes to bind to erythromycin. J.Mol.Biol. 48, 511-515.

TERAOKA, H., NIERHAUS, K.H. (1978) Ribosomal binding site of erythromycin. J.Mol.Biol. (in press)

TERAOKA, H., TANAKA, K. (1974) Properties of ribosomes from Streptomyces erythreus and Streptomyces griseus. J.Bacteriol. 120, 316-321

TERSHAK, D.R. (1976) Effect of hypertonic medium on protein synthesis of mahoney and LSc poliovirus. J.Virol. 20, 597-603

THIEMANN, J.E., CORONELLI, C., PAGANI, H., BERETTA, G., TAMONI, G., ARIOLI, V. (1968) Antibiotic production by new form-genera of actinomycetales. I. Sporangiomycin, an antibacterial agent isolated from Planomonospora parontospora var. antibiotica var. no. J.Antibiotics 21, 525-531

THOMPSON, M.R., STEINBERG, M.S., GEMSKI, P., FORMAL, S.B., DOCTOR, B.P. (1976) Inhibition of in vitro protein synthesis by Shigella disenteriae. Biochem.Biophys.Res.Commun. 71, 783-788

TIBONI, O., CIFERRI, O. (1971) Selective inhibition of the reactions catalyzed by ribosome-specific transfer factors G. FEBS Letters 19, 174-179

TIBONI, O., PARISI, B., CIFERRI, O. (1968) The mode of action of pederin, a drug inhibiting protein synthesis in eukaryotic organisms. Giornale Botanico Italiano 102, 337-345

TIMBERLAKE, W.E., GRIFFIN, D.H. (1974) Differential effects of cycloheximide and other inhibitors of protein synthesis on in vivo RNA synthesis in Achlya bisexualis. Biochim.Biophys.Acta 353, 248-252

TIMBERLAKE, W.E., McDOWELL, L., GRIFFIN, D.H. (1972) Cycloheximide inhibition of the DNA-dependent tRNA-polymerase I of Achlya bisexualis. Biochem.Biophys.Res.Commun. 46, 942-947

TOCCHINI-VALENTINI, G.P., MATTOCCIA, E. (1968) A mutant of E. coli with an altered supernatant factor. Proc.Nat.Acad.Sci. 61, 146-151

TOMLINSON, J.A., WALKER, V.M., FLEWETT, T.H., BACLAY, G.R. (1974) The inhibition of infection by cucumber mosaic virus and influence virus by extracts from Phytolocca americana. J.Gen. Virol. 22, 225-232

TOMPKINS, R.K., SCOLNICK, E.M., CASKEY, C.T. (1970) Peptide chain termination. VII. The ribosomal and release factors requirements for peptide release. Proc.Nat.Acad.Sci. 65, 702-708

TORI, K., TOKURA, K., OKABE, K., EBATA, K., OTSUKA, H., LUKACS, G. (1976) Tetrahedron Letters 3, 185-188

TOWERS, N.R., DIXON, H., KELLERMAN, G.M., LINNANE, A.W. (1972) Biogenesis of mitochondria. The sensitivity of rat liver mitochondria to antibiotics; a phylogenetic difference between a mammalian system and yeast. Arch.Biochem.Biophys. 151, 361-369

TOWERS, N.R., KELLERMAN, G.M., LINNANE, A.W. (1973) Competition between non-inhibitory antibiotics and inhibitory antibiotics for binding by rat liver mitochondrial ribosomes. Arch.Biochem. Biophys. 155, 159-166

TRAKATELLIS, A.C., MONTJAR, M., AXELROD, A.E. (1965) Effect of cycloheximide on polysomes and protein synthesis in the mouse liver. Biochemistry 4, 2065-2071

TRAUB, P., NOMURA, M. (1968) Streptomycin resistance mutation in Escherichia coli: altered ribosomal proteins. Science 160, 198-199

TRAUT, R.R., HEIMARK, R.L., SUN, T.T., HERSHEY, J.W.R., BOLLEN, A. (1974) In "Ribosomes" (Nomura, M., Tissieres, A. and Lengyel, P., eds.), pp. 271-308, Cold Spring Harbor: New York, USA

TRAUT, R.R., MONRO, R.E. (1964) The puromycin reaction and its relation to protein synthesis. J.Mol.Biol. 10, 63-72

TREOLAR, M.A., TREOLAR, M.E., KISSILEVSKY, R. (1977) Ethionine and the phosphorylation of ribosomal protein S6. J.Biol.Chem. 252, 6217-6221

TRITTON, T.T. (1977) Ribosome-tetracycline interactions. Biochemistry 16, 4133-4138

TSCHERNE, J.S., PESTKA, S. (1975) Inhibition of protein synthesis in intact HeLa cells. Ant.Agents Chemother. 8, 479-487

TULI, V. (1974) Mechanism of the antiviral action of 3-methyleneoxindole. Ant.Agents Chemother. 5, 485-491

TURNOWSKY, F., HÖGENAUER, G. (1973) Colicin E3, an inactivating agent of the ribosomal A-site. Biochem.Biophys.Res.Commun. 55, 1246-1254

UCHIDA, K., ICHIKAWA, T., SHIMAUCHI, Y., ISHIKURA, T., OZAKI, A. (1971) Hikizimycin, a new antibiotic. J.Antibiotics 24, 259-262

UCHIDA, K., WOLF, H. (1974) Metabolic products of microorganisms. 133. Inhibition of ribosomal peptidyl transferase by hikizimycin, a nucleoside antibiotic. J.Antibiotics 27, 783-787

UEHARA, Y., HORI, M., UMEZAWA, H. (1974) Negamycin inhibits termination of protein synthesis directed by phage f_2 RNA in vitro. Biochim. Biophys.Acta 374, 82-95

UEHARA, Y., HORI, M., UMEZAWA, H. (1976a) Specific inhibition of the termination process of protein synthesis by negamycin. Biochem. Biophys.Res.Commun. 442, 251-262

UEHARA, Y., HORI, M., UMEZAWA, H. (1976b) Inhibitory effect of negamycin on polysomal ribosomes of Escherichia coli. Biochim. Biophys.Acta 447, 406-412

UEHARA, Y., KONDO, S., UMEZAWA, H., SUZUKAKE, K., HORI, M. (1972) Negamycin, a miscoding antibiotic with a unique structure. J.Antibiotics 25, 685-688

UENO, Y. (1977) Mode of action of trichothecenes. Pure Applied Chem. 49, 1737-1745

UENO, Y., HOSOYA, M., MORITA, Y., UENO, I., TATSUNO, T. (1968) Inhibition of the protein synthesis in rabbit reticulocyte by nivalenol, a toxic principle isolated from Fusarium nivale-growing rice. J.Biochem. 64, 479-485

UENO, Y., UENO, I., TATSUNO, T., OHOKUBO, K., TSUNODA, H. (1969) Fusarenon-X, a toxic principle of Fusarium nivale-culture filtrate. Experientia 25, 1062

UMEZAWA, H. (1967) Index of Antibiotics from Actinomycin. University of Tokyo Press

UMEZAWA, H., OKAMI, Y., HASHIMOTO, T., SUHARA, Y., HAMADA, M., TAKEUCHI, T. (1965) A new antibiotic, kasugamycin. J.Antibiotics 18, 101-103

UMEZAWA, H., UEDA, M., MAEDA, K., YAGISHITA, K., KONDO, S., OKAMI, Y., UTAHARA, R., OSATO, Y., NITTA, K., TAKEUCHI, T. (1957) Production and isolation of a new antibiotic, kanamycin. J.Antibiotics 10, 181-188

UNO, H., OYABU, S., OHTSUKA, E., IKEHARA, M. (1971) The effect of guanosine 5'-triphosphate analogues on protein synthesis. Biochim.Biophys.Acta 228, 282-288

VAN de KLUNDERT, J.A.M., Den TURK, E., BORMAN, A.H., VAN der MEIDE, P.H., BOSCH, L. (1977) Isolation and characterization of a mocimycin resistant mutant of Escherichia coli with an altered elongation factor EF-Tu. FEBS Letters 81, 303-307

VAN DIEIJEN, G., VAN der LAKEN, C.J., VAN KNIPPENBERG, P.H., VAN DUIN, J. (1975) Function of Escherichia coli ribosomal protein S1 in translation of natural and synthetic messenger RNA. J.Mol. Biol. 93, 351-366

VAZQUEZ, D. (1963) Antibiotics which affect protein synthesis: the uptake of ^{14}C-chloramphenicol by bacteria. Biochem.Biophys.Res. Commun. 12, 409-413

VAZQUEZ, D. (1964a) The binding of chloramphenicol by ribosomes from Bacillus megaterium. Biochem.Biophys.Res.Commun. 15, 464-468

VAZQUEZ, D. (1964b) Uptake and binding of chloramphenicol by sensitive and resistant organisms. Nature 203, 257-260

VAZQUEZ, D. (1966a) Antibiotics affecting chloramphenicol uptake by bacteria. Their effect on amino acid incorporation in a cell-free system. Biochim.Biophys.Acta 114, 289-295

VAZQUEZ, D. (1966b) Binding of chloramphenicol to ribosomes. The effect of a number of antibiotics. Biochim.Biophys.Acta 114, 277-288

VAZQUEZ, D. (1966c) Studies on the mode of action of the streptogramin antibiotics. J.Gen.Microbiol. 42, 93-106

VAZQUEZ, D. (1967a) Inhibitors of protein synthesis at the ribosome level. Studies on their site of action. Life Sciences 6, 381-386

VAZQUEZ, D. (1967b) Binding to ribosomes and inhibitory effect on protein synthesis of the spiramycin antibiotics. Life Sciences 6, 845-853

VAZQUEZ, D. (1974) Inhibitors of protein synthesis. FEBS Letters 40, S63-S84 (Suppl.)

VAZQUEZ, D. (1975a) In "Antibiotics. III. Mechanism of Action of Antimicrobial and Antitumor Agents" (Corcoran, J.W. and Hahn, F.E., eds.), pp. 459-479, Springer-Verlag: Berlin-Heidelberg-New York

VAZQUEZ, D. (1975b) In "Antibiotics. III. Mechanism of Action of Antimicrobial and Antitumor Agents" (Corcoran, J.W. and Hahn, F.E., eds.), pp. 521-534, Springer-Verlag: Berlin-Heidelberg-New York

VAZQUEZ, D. (1978a) Translation inhibitors. In "Amino Acid and Protein Biosynthesis II. International Review" (Arnstein, H.R.V., ed.), Vol. 18, University Park Press: Baltimore

VAZQUEZ, D. (1978b) Protein and glycoprotein toxins that inactivate the eukaryotic ribosome. In "Antibiotics" (Hahn, F.E., ed.), Vol. V, Springer-Verlag: Berlin-Heidelberg-New York (in press)

VAZQUEZ, D., BATTANER, E., NETH, R., HELLER, G., MONRO, R.E. (1969) The function of 80S ribosomal subunits and effects of some inhibitors. Cold Spring Harb.Symp.Quant.Biol. 34, 369-375

VAZQUEZ, D., CONTRERAS, A., FRESNO, M., JIMENEZ, A. (1977) Interaction of translation inhibitors with ribosomes and polysomes. In "Translation of Natural and Synthetic Polynucleotides".Proceedings of the International Conference on Translation of Natural and Synthetic Polynucleotides (Legocki, A.B., ed.), pp. 380-385, University of Agriculture in Poznan

VAZQUEZ, D., MONRO, R.E. (1967) Effects of some inhibitors of protein synthesis on the binding of aminoacyl-tRNA to ribosomal subunits. Biochim.Biophys.Acta 142, 155-173

VERNIE, L.N., BONT, W.S., GINJAAR, H.B., EMMELOT, P. (1975) Elongation factor 2 as the target of the reaction product between sodium selenite and glutathione (GSSeSG) in the inhibiting of amino acid incorporation in vitro. Biochim.Biophys. Acta 414, 238-292

VESCO, C., COLOMBO, B. (1970) Effect of sodium fluoride on protein synthesis in HeLa cells: inhibition of ribosome dissociation. J.Mol.Biol. 47, 335-352

VIAU, J.P., DAVIS, F.F. (1970) Effect of cycloheximide on the synthesis and modification of ribosomal RNA in Neurospora crassa. Biochim.Biophys.Acta 209, 190-195

VILLAFRANCA, J.E., ROBERTUS, J.D. (1977) Crystallographyc study of the anti-tumor protein ricin. J.Mol.Biol. 116, 331-335

VINCE, R., WEISS, D., GORDON, J., HOWARD, G., SMITH, I., PESTKA, S. (1976) Binding of thiostrepton to ribosomes from thiostrepton-sensitive and -resistant Bacillus subtilis strains. Ant.Agents Chemother. 9, 665-667

VOGEL, Z., VOGEL, T., ELSON, D. (1971) The effect of erythromycin on peptide bond formation and the termination reaction. FEBS Letters 15, 249-253

VOGEL, Z., VOGEL, T., ZAMIR, A., ELSON, D. (1970) Ribosome activation and the binding of dihydrostreptomycin: effect of polynucleotides and temperature of activation. J.Mol.Biol. 54, 379-386

VOS, C., VERWIEL, P.E.J. (1973) The total structure of the novel antibiotic mocimycin (MYC 8003). Tetrahedron Letters 52, 5173-5176

VUILHORGNE, M., ENNIFAR, S., BHUPESH, C.D., PASCHAL, J.W., NAGARAJAN, R., HAGAMAN, E.W., WENKERT, E. (1977) Structure analysis of the nucleoside disaccharide antibiotic anthelmycin by carbon-13 nuclear magnetic resonance spectroscopy. A structural revision of hikizimycin and its identity with anthelmycin. J.Org.Chem. 42, 3289-3291

WACKER, A., ISHIMOTO, M., CHANDRA, P. (1967) Spezifische und unspezifische hemmung der zellfreien proteinsynthese mit polyanionen. Zeitschrift Naturforschung 22, 413-417

WAGNER, E., SCHWEIGER, M., PONTA, H., HERRLICH, P. (1977) Messenger-selective inhibitor for the initiation of translation in Escherichia coli: nitrofurantoin. FEBS Letters 83, 337-340

WAISVISZ, J.M., VAN der HOEVEN, M.G., VAN PEPPEN, J., ZWENNIS, W.C.M. (1957) Bottromycin. I. A new sulfur-containing antibiotic. J.Amer.Chem.Soc. 79, 4520-4522

WAKSMAN, S.A., LECHEVALIER, H. (1949) Neomycin, a new antibiotic active against streptomycin-resistant bacteria including tuberculosis organisms. Science 109, 305

WALDMAN, A.A., GOLDSTEIN, J. (1973) Rabbit reticulocyte ribosomes and Krebs ascites supernatant: a mixed system dependent upon added mRNA. Biochim.Biophys.Acta 331, 243-250

WALDMAN, A.A., MARX, G., GOLDSTEIN, D. (1974) Heparin as inhibitor of mammalian protein synthesis. II. Degree of sulfation; related sulfated mucopolysaccharides. Biochim.Biophys.Acta 343, 324-329

WALDMAN,A.A., MARX, G., GOLDSTEIN, J. (1975) Isolation of rabbit reticulocyte initiation factors by means of heparin bound to sepharose. Proc.Nat.Acad.Sci. 72, 2352-2356

WALLACE, B.J., DAVIS, B.D. (1973) Cyclic blockade of initiation sites by streptomycin-damaged ribosomes in Escherichia coli: an explanation for dominance of sensitivity. J.Mol.Biol. 73, 377-390

WALLACE, B.J., TAI, P.C., DAVIS, B.D. (1973) Effect of streptomycin on the response of Escherichia coli ribosomes to the dissociation factor. J.Mol.Biol. 75, 391-400

WALLACE, B.J., TAI, P.C., DAVIS, B.D. (1974) Selective inhibition of initiating ribosomes by spectinomycin. Proc.Nat.Acad.Sci. 71, 1634-1638

WALLACE, B.J., TAI, P.C., HERZOG, E.L., DAVIS, B.D. (1973) Partial inhibition of polysomal ribosomes of Escherichia coli by streptomycin. Proc.Nat.Acad.Sci. 70, 1234-1237

WALLER, C.W., FRYTH, P.W., HUTCHINGS, B.L., WILLIAMS, J.H. (1953) Achromycin. The structure of the antibiotic puromycin. J.Amer. Chem.Soc. 75, 2025

WALLER, J.P., ERDOS, T., LEMOINE, F., GUTTMANN, S., SANDRIN, E. (1966) Inhibition of protein synthesis by aminoacyl 3'(2')-adenosine. Biochim.Biophys.Acta 119, 566-580

WATANABE, S. (1972) Interaction of siomycin with the acceptor site of Escherichia coli. J.Mol.Biol. 67, 443-457

WAX, R., MAIESE, W., WASTON, R., BIRNBAUM, J. (1976) Efrotomycin a new antibiotic from Streptomyces lactamdurans. J.Antibiotics 29, 670-673

WEBER, L.A., HICKEY, E.D., MARONEY, P.A., BAGLIONI, C. (1977) Inhibition of protein synthesis by Cl^-. J.Biol.Chem. 252, 4007-4101

WEEKS, D.P., BAXTER, R. (1972) Specific inhibition of peptide-chain initiation by 2-(4-methyl-2,6-dinitroanilino)-N-methylpropionamide. Biochemistry 11, 3060-3064

WEI, C.M., CAMPBELL, I.M., McLAUGHLIN, C.S., VAUGHAN, M.H. (1974) Binding of trichodermin to mammalian ribosomes and its inhibition by other 12,13-epoxytrichothecenes. Mol.Gen.Biochem. 3, 215-219

WEINSTEIN, I.B., FRIEDMAN, S.M., OCHOA, M. (1966) Fidelity during translation of the genetic code- Cold Spring Harb.Symp.Quant. Biol. 31, 671-681

WEINSTEIN, M.J., LUEDEMANN, G.H., ODEN, E.M., WAGMAN, G.H. (1964) Gentamicin, a new broad-spectrum antibiotic complex. Ant.Agents Chemother. 1963, 1-7

WEINSTEIN, M.J., LUEDEMANN, G.H., ODEN, E.M., WAGMAN, G.H., ROSSELET, J.P., MARQUEZ, J.A., CONIGLIO, C.T., CHARNEY, W., HERZOG, H.L., BLACK, J. (1963) Gentamicin, a new antibiotic complex from Micromonospora. J.Med.Chem. 6, 463-464

WEISBLUM, B., DAVIES, J. (1968) Antibiotic inhibitors of the bacterial ribosome. Bacteriol.Rev. 32, 493-528

WEISBLUM, B., DEMOHN, V. (1970) Thiostrepton, an inhibitor of 50S ribosome subunit function. J.Bacteriol. 101, 1073-1075

WEISSBACH, H., BROT, N. (1970) Inhibition of transfer factor Ts by aurintricarboxylic acid. Biochem.Biophys.Res.Commun. 39, 1194-1198

WEISSBACH, H., PESTKA, S. (1977) Molecular Mechanisms of Protein Synthesis. Academic Press: New York-San Francisco-London

WEISSBACH, H., REDFIELD, B., YAMASAKI, E., DAVIS, R.C., PESTKA, S., BROT, N. (1972) Studies on the ribosomal sites involved in factors Tu and G-dependent reactions. Arch.Biochem.Biophys. 149, 110-117

WENGLER, G., WENGLER, G. (1972) Medium hypertonicity and polyribosome structure in HeLa cells. The influence of hypertonicity of the growth medium on polyribosomes in HeLa cells. Eur.J.Biochem. 27, 162-173

WETTSTEIN, F.P., NOLL, H., PENMAN, S. (1964) Effect of cycloheximide on ribosomal aggregates engaged in protein synthesis *in vitro*. Biochim.Biophys.Acta 87, 525-528

WHIFFEN, A.J., BOHONOS, N., EMERSON, R.L. (1946) The production of antifungal antibiotic by *Streptomyces griseus*. J.Bacteriol. 52, 610-611

WHITE, J.P., CANTOR, C.R. (1971) Role of magnesium in the binding of tetracycline to *Escherichia coli* ribosomes. J.Mol.Biol. 58, 397-400

WICK, W.E., WELLES, J.S. (1968) Nebramycin, a new broad-spectrum antibiotic complex. IV In vitro and in vivo laboratory evaluation. Ant.Agents Chemother. 1967, 332-340

WIDELITZ, M.M., CORYELL, M.R., AVADHANI, N.G. (1977) Reversal of amphetamine induced polysome dissociation by neuroleptic agents in rat brain. Biochem.Biophys.Res.Commun. 76, 1223-1229

WIENEN, B., STOFFLER, G., SMITH, I., WEISS, D., VINCE, R., PESTKA, S. (1978) Ribosomal protein alterations in thiostrepton- and micrococcin-resistant mutants of *Bacillus subtilis*. Mol.Gen. Genetic (in press)

WILDMAN, W.C. (1968) in "The Alkaloids" (Manske, R.H.F., ed.), Vol. 11, Chapter 10, pp. 307-405, Academic Press: New York.

WILEY, P.F., MacKELLAR, F.A. (1970) The structure of sparsomycin. J.Amer.Chem.Soc. 92, 417-418

WILHELM, J.M., OLEINICK, N.L., CORCORAN, J.W. (1968) Interaction of antibiotics with ribosomes: structure-function relationships and a possible common mechanism for the antibacterial action of the macrolides and lincomycin. Ant.Agents Chemother. 1967, 236-250

WILHELM, J.M., PETTIT, S.E., JESSOP, J.J. (1978) Aminoglycoside antibiotics and eukaryotic protein synthesis: structure-function relationship in the stimulation of misreading with a wheat embryo system. Biochemistry 17, 1143-1149

WILLIE, G.R., RICHMAN, N., GODFREDSEN, W.O., BODLEY, J.W. (1975) Some characteristics of an structural requirements for the interaction of 24, 25 dihydrofusidic acid with ribosome elongation factor G complexes. Biochemistry 14, 1713-1718

WITTMANN, H.G., APIRION, D. (1975) Analysis of ribosomal proteins in streptomycin resistant and dependent mutants isolated from streptomycin independent Escherichia coli strains. Mol.Gen. Genetics 141, 331-341

WITTMANN, H.G., STOFFLER, G., APIRION, D., ROSEN, L., TANAKA, K., TAMAKI, M., TAKATA, R., DEKIO, S., OTAKA, E., OSAWA, S. (1973) Biochemical and genetic studies on two different types of erythromycin resistant mutants of Escherichia coli with altered ribosomal proteins. Mol.Gen.Genetics 127, 175-189

WOLF, H. (1973) Avilamycin, an inhibitor of the 30S ribosomal subunit function. FEBS Letters 36, 181-186

WOLF, H., CHINALI, G., PARMEGGIANI, A. (1974) Kirromycin, an inhibitor of protein synthesis that acts on elongation factor Tu. Proc. Nat.Acad.Sci. 71, 4910-4914

WOLF, H., CHINALI, G., PARMEGGIANI, A. (1977) Mechanism of the inhibition of protein synthesis by kirromycin. Role of elongation factor Tu and ribosomes. Eur.J.Biochem. 75, 67-75

WOLF, H., ZÄHNER, H. (1972) Stoffwechselprodukte von Mikroorganismen 99. Mitteilung. Kirromycin. Arch.Mikrobiol. 83, 147-154

WOLF, H., ZÄHNER, H., NIERHAUS, K. (1972) Kirromycin, an inhibitor of the 30S ribosomal subunit function. FEBS Letters 21, 347-350

WOLFE, A.D., HAHN, F.E. (1964) Erythromycin: mode of action. Science 143, 1445-1446

WOLFE, A.D., HAHN, F.E. (1965) Mode of action of chloramphenicol. IX. Effects of chloramphenicol upon a ribosomal amino acid polymerization system and its binding to bacterial ribosome. Biochim.Biophys.Acta 95, 146-155

WOLFGANG, R.W., LAWRENCE, N.L. (1967) Binding of streptomycin by ribosomes of sensitive, resistant and dependent Bacillus megaterium. J.Mol.Biol. 29, 531-535

WOO, P.W.K., DION, H.W., BARTZ, Q.R. (1971) Butirosins A and B, aminoglycoside antibiotics. III. Structures. Tetrahedron Letters 1971, 2625-2628

WU, J.M., CHEUNG, Ch.P., SUHADOLNIK, R.J. (1977) Differential inhibition with partially purified and endogenous rabbit reticulocyte globin mRNA by 7-methylguanosine 5'-monophosphate. Biochem.Biophys.Res.Commun. 78, 1079-1086

WYATT, S., SHEPHERD, R. (1969) Isolation and characterization of a virus inhibitor from Phytolacca americana. Phytopathology 59, 1787-1794

YAGUCHI, M., WITTMANN, H.G. (1976) Alteration of ribosomal protein S17 by mutation linked to neamine resistance in Escherichia coli. II. Localization of the amino acid replacement in protein S17 from a nea A mutant. J.Mol.Biol. 104, 617-620

YAMADA, T., MASUDA, K., MIZUGUCHI, Y., SUGA, K. (1976) Altered ribosomes in antibiotic-resistant mutants of Mycobacterium smegmatis. Ant.Agents Chemother. 9, 817-823

YAMADA, T., MASUDA, K., SHOJI, K., HORI, M. (1972) Analysis of ribosomes from viomycin-sensitive and resistant strains of Mycobacterium smegmatis. J.Bacteriol. 112, 1-6

YAMADA, T., MASUDA, K., SHOJI, K., HORI, M. (1974) Pleiotropic antibiotic resistance mutations associated with ribosomes and ribosomal subunits in Mycobacterium smegmatis. Ant.Agents Chemother. 6, 46-53

YAMAGUCHI, H. (1961) Action mechanism of mikamycins. I. Effect of mikamycins on protein and nucleic acid metabolism. J.Antibiotics 14, 313-323

YAMAGUCHI, H., NOKAYAMA, Y., TAKEDA, K., TAWARA, K. (1957) A new antibiotic, althiomycin. J.Antibiotics 10, 195-200

YAMAGUCHI, H., TANAKA, N. (1964) Selective toxicity of mikamycins inhibitors of protein synthesis. Nature 201, 499-501

YAMAGUCHI, H., TANAKA, N. (1966) Inhibition of protein synthesis by blasticidin S. II. Studies on the site of action in E. coli polypeptide synthesizing systems. J.Biochem. 60, 632-642

YAMAGUCHI, H., YAMAMOTO, C., TANAKA, N. (1965) Inhibition of protein synthesis by blasticidin S. I. Studies with cell-free systems from bacterial and mammalian cells. J.Biochem. 57, 667-677

YAMAKI, H., NISHIMURA, T., KUBOTA, K., KINOSHITA, T., TANAKA, N. (1974) Selective inhibition of initiation of globin synthesis by phenomycin. Biochem.Biophys.Res.Commun. 59, 482-488

YARMOLINSKY, M.B., HABA, G. de la. (1959) Inhibition by puromycin of amino acid incorporation into protein. Proc.Nat.Acad.Sci. 45, 1721-1729

YOKOSAWA, H., INOUE-YOKOSAWA, N., ARAI, K., KAWAKITA, M., KAZIRO, Y. (1973) The role of GTP hydrolysis in elongation factor Tu promoted binding of aminoacyl-tRNA to ribosomes. J.Biol.Chem. 248, 375-377

YOKOZAWA, H., KAWAKITA, M., ARAI, K., INOUE-YOKOSAWA, N., KAZIRO, Y. (1975) Binding of aminoacyl-tRNA to ribosomes promoted by elongation factor Tu. J.Biochem. 77, 719-728

YONEHARA, H., OTAKE, N. (1966) Absolute configuration of blasticidin S. Tetrahedron Letters, pp. 3785-3792

ZAMIR, A., LEDER, P., ELSON, D. (1966) A ribosome-catalyzed reaction between N-formylmethionyl-tRNA and puromycin. Proc. Nat.Acad.Sci. 56, 1794-1801

ZEMLICKA, J., CHLADEK, S., RINGER, D., QUIGGLE, K. (1975) Substrate specificity of ribosomal peptidyltransferase. Effect of modification in the heterocyclic carbohydrate and amino acid moiety of 2'(3')-O-L-phenyladenosine. Biochemistry 14, 5239-5249

ZIMMERMANN, R.A., IKEYA, Y., SPARLING, F. (1973) Alteration of ribosomal protein S4 by mutation linked to kasugamycin-resistance in Escherichia coli. Proc.Nat.Acad.Sci. 70, 71-75

Subject Index

The underlined page numbers refer to main entries (chapters).

Molecular Biology Biochemistry and Biophysics

Editors: A. Kleinzeller, G. F. Springer, H. G. Wittmann

Volume 20: M. A. Lauffer

Entropy – Driven Processes in Biology

Polymerization of Tobacco Mosaic Virus Protein and Similar Reactions
1975. 90 figures. X, 264 pages
ISBN 3-540-06933-X

This book deals with the nature of entropy-driven processes and the role of water. The phenomena examined are the polymerization of tobacco mosaic virus protein and many other proteins, the sickling phenomenon, the formation of structural and contractile elements, and cellular processes, such as pseudopodia formation and spindle formation, which depend upon the polymerization of tubulin.

Volume 21: R. C. Burns, R. W. F. Hardy

Nitrogen Fixation in Bacteria and Higher Plants

1975. 27 figures. X, 189 pages
ISBN 3-540-07192-X

This detailed exposition of the biochemistry and biophysics on N_2 fixation is supplemented by discussion of relevant historical, taxonomic, physiological, and evolutionary material. The author's aim is to provide an up-to-date description of existing knowledge and an evaluation of current work in this area. The biomechanical and biophysical aspects of N_2 fixation are explored in depth at specialist level, yet the general science reader is given a comprehensible overview of these topics so that he can place them in appropriate temporal and ecological perspective. This is the first time nitrogen fixation has been given such breadth of coverage in a single work and with the continuity possible only in a monograph.

Volume 22: H. J. Fromm

Initial Rate Enzyme Kinetics

1975. 88 figures, 19 tables. X, 321 pages
ISBN 3-540-07375-2

This book fills a gap in the literature on enzyme kinetics. It outlines how one may reach definitive conclusions regarding enzyme and substrate interaction, the functional groups of enzymes involved in catalysis, mapping of the active site, intermediates involved in enzyme catalysis, allosteric mechanisms and evaluation of kinetic parameters that are essential for an understanding of enzyme regulation. The text covers such topics as kinetic nomenclature, systematic procedures for the derivation of rate equations, reversible enzyme inhibition, isotope exchange, temperature and pH effects on catalysis, and allostery. The experimental protocol involved in many kinetic procedures is discussed in detail. Specific examples from the literature are used to illustrate each topic.

Volume 23: M. Luckner, L. Nover, H. Böhm

Secondary Metabolism and Cell Differentiation

1977. 52 figures, 7 tables. VI, 130 pages
ISBN 3-540-08081-3

The volume is based on the concept that the biosynthesis of secondary natural substances is a widely found characteristic of cell specialization in almost all organisms. The central theme together with chapters on coordinated enzyme synthesis and possible effectors of gene expression in this field of metabolism, is the analysis of differentiating programs, including the formation of enzymes of secondary metabolism. Particular aspects of the formation of secondary natural substances in plant cell culture are treated in a separate chapter.

Volume 24

Chemical Relaxation in Molecular Biology

Editors: I. Pecht, R. Rigler
1977. 141 figures, 50 tables. XVI, 418 pages
ISBN 3-540-08173-9

The purpose of this monograph is to give a representative cross section of the current research activities dedicated to the analysis of essential steps in biological reactions. This covers the range of the following topics: hydrogen-bond formation, nucleotide base pairing, protein folding, isomerisation of protein and nucleic acid conformations, interactions between protein and proteins, nucleic acid and proteins, enzymes and substrates, antibody and haptens or ionic transport through membranes. A common denominator in these studies is the search for an understanding of the laws that govern the dynamic behaviour of living systems.

Volume 25

Advanced Methods in Protein Sequence Determination

Editor: S.B. Needleman
1977. 97 figures, 25 tables. XII, 189 pages
ISBN 3-540-08368-5

The determination of protein sequences has become so commonplace that, as more laboratories have entered this area of study, the sophistication of the tecnology has, in fact, progressed to make use of physical properties not previously utilized for this purpose. Earlier manual techniques have become automated; current instrumentation operates at higher parameters and with greater precision than before. Thus the present volume supplements the earlier one in presenting details of the more advanced technologies (optical, high pressure, X-ray, immunology etc.) being used in sequence determination today.

Volume 26
A.S. Brill

Transition Metals in Biochemistry

1977. 49 figures, 18 tables. VIII, 186 pages
ISBN 3-540-08291-3

This monograph is a concise review of the current state of and developments in the field, with emphasis upon the application of physical methods to the investigation of metal coordination. Molecular functions of proteins containing transition metal ion prosthetic groups are summarized. Where established by X-ray diffraction, the three-dimensional structures of relevant metal binding sites in proteins are described. Light absorption and electron paramagnetic resonance are treated in depth. Those aspects of the theory are presented which can be directly employed in the quantification and interpretation of experimental data. The monograph provides a basis for closer communication between scientists of different backgrounds with a common interest in the biochemistry of transition metal ions.

Volume 27

Effects of Ionizing Radiation on DNA

Physical, Chemical and Biological Aspects
Editors: A.J. Bertinchamps (Coordinating Editor), J. Hüttermann, W. Köhnlein, R. Téoule
1978. 74 figures, 48 tables. XXII, 383 pages
ISBN 3-540-08542-4

For the first time, the three essential approaches to research on the effects of ionizing radiation on DNA and its constituents have been described together in one book, providing an overall view of the fundamental problems involved. A result of the European study group on "Primary Effects of Radiation on Nucleic Acids", this book contains the current state of knowledge in this field, and has been written in close collaboration by 27 authors.

Volume 28
A. Levitzki

Quantitative Aspects of Allosteric Mechanisms

1978. 13 figures, 2 tables. VIII, 106 pages
ISBN 3-540-08696-X

This book provides a concise but comprehensive treatment of the basic regulatory phenomena of allostery and cooperativity. It critically evaluates the differences between the allosteric models and their applicability to real situations. For the first time the full analysis of the different allosteric models is given, and compared with the pure thermodynamic approach. The treatment of the subject of allostery in this book is of great value to enzymologists, receptorologists, pharmacologists and endocrinologists, as it provides the basic rules for the study of ligand-protein and ligand-receptor interactions.

Volume 29
E. Heinz

Mechanics and Energetics of Biological Transport

1978. 35 figures, 3 tables. XV, 159 pages
ISBN 3-540-08905-5

This book presents the interrellations of mechanistic models on the one hand and the kinetic and energetic behavior of transport and permeatin processes on the other, using the principles of irreversible thermodynamics. The advantages of each method are compared. The special aim is to show how to appropriate formulas can be transformed into each other, in order to recognize in what way the kinetic parameters correspond to those of irreversible thermodynamics.

Springer-Verlag
Berlin
Heidelberg
New York

MIX
Papier aus verantwortungsvollen Quellen
Paper from responsible sources
FSC® C105338

FSC
www.fsc.org

If you have any concerns about our products,
you can contact us on
ProductSafety@springernature.com

In case Publisher is established outside the EU,
the EU authorized representative is:
Springer Nature Customer Service Center GmbH
Europaplatz 3, 69115 Heidelberg, Germany

Printed by Libri Plureos GmbH
in Hamburg, Germany